Edge Intelligence in the Making

Optimization, Deep Learning, and Applications

Synthesis Lectures on Learning, Networks, and Algorithms

Editor
Lei Ying, *University of Michigan, Ann Arbor*

Editor Emeritus
R. Srikant, *University of Illinois at Urbana-Champaign*

Founding Editor Emeritus
Jean Walrand, *University of California, Berkeley*

Synthesis Lectures on Learning, Networks, and Algorithms is an ongoing series of 75- to 150-page publications on topics on the design, analysis, and management of complex networked systems using tools from control, communications, learning, optimization, and stochastic analysis. Each lecture is a self-contained presentation of one topic by a leading expert. The topics include learning, networks, and algorithms, and cover a broad spectrum of applications to networked systems including communication networks, data-center networks, social, and transportation networks. The series is designed to:

- Provide the best available presentations of important aspects of complex networked systems.

- Help engineers and advanced students keep up with recent developments in a rapidly evolving field of science and technology.

- Facilitate the development of courses in this field.

Energy-Efficient Scheduling under Delay Constraints for Wireless Networks
Randall Berry, Eytan Modiano, and Murtaza Zafer
2012

NS Simulator for Beginners
Eitan Altman and Tania Jiménez
2012

Network Games: Theory, Models, and Dynamics
Ishai Menache and Asuman Ozdaglar
2011

An Introduction to Models of Online Peer-to-Peer Social Networking
George Kesidis
2010

Stochastic Network Optimization with Application to Communication and Queueing
Systems
Michael J. Neely
2010

Scheduling and Congestion Control for Wireless and Processing Networks
Libin Jiang and Jean Walrand
2010

Performance Modeling of Communication Networks with Markov Chains
Jeonghoon Mo
2010

Communication Networks: A Concise Introduction
Jean Walrand and Shyam Parekh
2010

Path Problems in Networks
John S. Baras and George Theodorakopoulos
2010

Performance Modeling, Loss Networks, and Statistical Multiplexing
Ravi R. Mazumdar
2009

Network Simulation
Richard M. Fujimoto, Kalyan S. Perumalla, and George F. Riley
2006

Edge Intelligence in the Making: Optimization, Deep Learning, and Applications
Sen Lin, Zhi Zhou, Zhaofeng Zhang, Xu Chen, and Junshan Zhang

ISBN: 978-3-031-01252-5   paperback
ISBN: 978-3-031-02380-4   ebook
ISBN: 978-3-031-00244-1   hardcover

DOI 10.1007/978-3-031-02380-4

A Publication in the Springer Nature series
SYNTHESIS LECTURES ON ADVANCES IN AUTOMOTIVE TECHNOLOGY

Lecture #25
Editor: Lei Ying, *University of Michigan, Ann Arbor*
Editor Emeritus: R. Srikant, *University of Illinois at Urbana-Champaign*
Founding Editor Emeritus: Jean Walrand, *University of California, Berkeley*
Series ISSN
Print 2690-4306   Electronic 2690-4314

Edge Intelligence in the Making

Optimization, Deep Learning, and Applications

Sen Lin
Arizona State University, Tempe, AZ

Zhi Zhou
Sun Yat-sen University, Guangzhou, China

Zhaofeng Zhang
Arizona State University, Tempe, AZ

Xu Chen
Sun Yat-sen University, Guangzhou, China

Junshan Zhang
Arizona State University, Tempe, AZ

SYNTHESIS LECTURES ON LEARNING, NETWORKS, AND ALGORITHMS
#25

ABSTRACT

With the explosive growth of mobile computing and Internet of Things (IoT) applications, as exemplified by AR/VR, smart city, and video/audio surveillance, billions of mobile and IoT devices are being connected to the Internet, generating zillions of bytes of data at the network edge. Driven by this trend, there is an urgent need to push the frontiers of artificial intelligence (AI) to the network edge to fully unleash the potential of IoT big data. Indeed, the marriage of edge computing and AI has resulted in innovative solutions, namely edge intelligence or edge AI. Nevertheless, research and practice on this emerging inter-disciplinary field is still in its infancy stage. To facilitate the dissemination of the recent advances in edge intelligence in both academia and industry, this book conducts a comprehensive and detailed survey of the recent research efforts and also showcases the authors' own research progress on edge intelligence. Specifically, the book first reviews the background and present motivation for AI running at the network edge. Next, it provides an overview of the overarching architectures, frameworks, and emerging key technologies for deep learning models toward training/inference at the network edge. To illustrate the research problems for edge intelligence, the book also showcases four of the authors' own research projects on edge intelligence, ranging from rigorous theoretical analysis to studies based on realistic implementation. Finally, it discusses the applications, marketplace, and future research opportunities of edge intelligence. This emerging interdisciplinary field offers many open problems and yet also tremendous opportunities, and this book only touches the tip of iceberg. Hopefully, this book will elicit escalating attention, stimulate fruitful discussions, and open new directions on edge intelligence.

KEYWORDS

edge intelligence, edge computing, artificial intelligence, deep learning, Internet of Things (IOT), model training, model inference

Contents

Preface

With the breakthroughs in deep learning, recent years have witnessed a boom of artificial intelligence (AI) applications and services, spanning from personal assistant to recommendation systems to video/audio surveillance. More recently, with the proliferation of mobile computing and Internet of Things (IoT), billions of mobile and IoT devices are connected to the Internet, generating zillions of bytes of data at the network edge. Driven by this trend, there is an urgent need to push the AI frontiers to the network edge so as to fully unleash the potential of the edge big data. To meet this demand, edge computing—an emerging paradigm that pushes computing tasks and services from the network core to the network edge—has been widely recognized as a promising solution. The resulting interdisciplinary area, namely edge intelligence (or edge AI), is beginning to garner a tremendous amount of interest. Nevertheless, research on edge intelligence is still in its infancy stage, and a dedicated venue for exchanging the recent advances of edge intelligence is highly desired by both the computer system and AI communities.

To this end, we conduct a comprehensive and detailed survey of the recent research efforts and also showcase our own research progress on edge intelligence in this book. Specifically, we first review the background and motivation for AI running at the network edge. Next, we provide an overview of the overarching architectures, frameworks, and emerging key technologies for deep learning modeled toward training/inference at the network edge. To better illustrate the research problems for edge intelligence, we also showcase four of our own research projects on edge intelligence, ranging from rigorous theoretical analysis to studies based on realistic implementation. Finally, we discuss the applications, marketplace, and future research opportunities on edge intelligence. We hope that this book will elicit escalating attention, stimulate fruitful discussions, and inspire further research ideas on edge intelligence.

Sen Lin, Zhi Zhou, Zhaofeng Zhang, Xu Chen, and Junshan Zhang
October 2020

Acknowledgments

The work of Junshan Zhang, Sen Lin, and Zhaofeng Zhang is supported in part by the NSF under Grants SaTC-1618768 and CPS-1739344 and the ARO under grant W911NF-16-1-0448; and the work of Xu Chen and Zhi Zhou is supported in part by the NSFC under grants U1711265 and 61972432 and the GDIIET under grant 2017ZT07X355.

Sen Lin, Zhi Zhou, Zhaofeng Zhang, Xu Chen, and Junshan Zhang
October 2020

CHAPTER 1

Introduction to Edge Intelligence

We are living during an unprecedented era of artificial intelligence (AI) expansion. Driven by the recent advancements of algorithms, computing power, and big data, deep learning [Lecun et al., 2015]—the most dazzling sector of AI—has made substantial breakthroughs in a wide spectrum of fields, ranging from computer vision, speech recognition, and natural language processing to chess playing (e.g., AlphaGo) and robotics [Deng et al., 2014]. Benefiting from these breakthroughs is a set of intelligent applications, as exemplified by intelligent personal assistants, personalized shopping recommendation, video surveillance, and smart home appliances, which have quickly ascended to the spotlight and gained enormous popularity. It is widely recognized that these intelligent applications are significantly enriching people's lifes, improving human productivity and enhancing social efficiency.

As a key driver that boosts AI development, big data has recently gone through a radical shift of data source from the mega-scale cloud datacenters to the increasingly widespread end devices, e.g., mobile devices and Internet of Things (IoT) devices. Traditionally, big data, such as online shopping records, social media contents, and business informatics, were mainly born and stored at mega-scale datacenters. However, with the proliferation of mobile computing and IoT, the trend is descending to the network edge. Specifically, Cisco estimates that nearly 850 ZB will be generated by all people, machines, and things at the network edge by 2021 [Cisco, 2016]. In sharp contrast, the global datacenter traffic will only reach 20.6 ZB by 2021. Clearly, via machine learning over the huge volumes of data at the edge devices, the edge ecosystem has great potential to create many novel application scenarios for AI and fuel the continuous boom of AI.

Pushing the AI frontier to the edge ecosystem that resides at the last mile of the Internet, however, is highly non-trivial, due to concerns of performance, cost, and privacy. Toward this goal, the conventional wisdom is to transport the data bulks from the IoT devices to the cloud datacenters for analytics [Heintz et al., 2015]. However, when moving a tremendous amount of data across the wide-area-network (WAN), both monetary cost and transmission delay can be prohibitively high, and the privacy leakage can also be a major concern [Pu et al., 2015]. An alternative is on-device analytics that run AI applications on the device to process the IoT data locally, which, however, may suffer from poor performance and energy efficiency. This is because

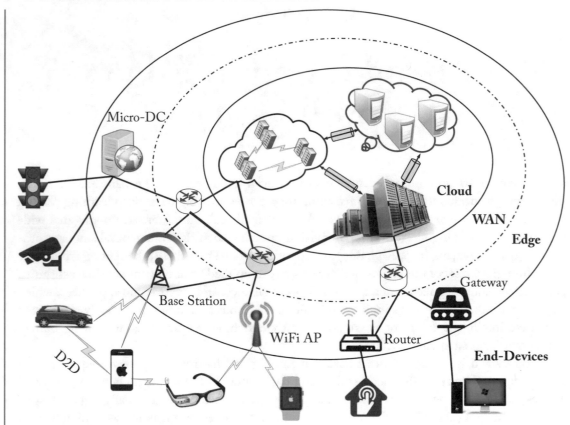

Figure 1.1: An illustration of edge computing.

many AI applications require high computational power that greatly outweigh the capacity of resource- and energy-constrained IoT devices.

To address the above challenges, edge computing (see Shi et al. [2016] and the references therein) has recently been proposed, which pushes cloud services from the network core to the network edges that are in closer proximity to IoT devices and data sources. We caution that it is not a binary choice between Cloud and Edge: they form a mutually beneficial, inter-dependent continuum. As illustrated in Figure 1.1, an edge node can be nearby end-device connectable by device-to-device (D2D) communications [Chen et al., 2017a], a server attached to an access point (e.g., WiFi, router, base station), a network gateway, or even a micro-datacenter available for use by nearby devices. While edge nodes can be varied in size: ranging from a credit-card-sized computer to a micro-datacenter with several server racks, physical proximity to the information-generation sources is the most crucial characteristic emphasized by edge computing. Essentially, the physical proximity between the computing and information-generation sources promises several benefits compared to the traditional cloud-based computing paradigm, including low-latency, energy-efficiency, privacy protection, reduced bandwidth consumption, on-premises and context-awareness [Mao et al., 2017c, Shi et al., 2016].

Indeed, the marriage of edge computing and AI has resulted in innovative solutions, namely "edge intelligence" or "edge AI" [Li et al., 2018b, Wang et al., 2019b]. Instead of relying on the cloud, edge intelligence makes use of most of the widespread edge resources to gain AI insight. Notably, edge intelligence has garnered much attention from both the industry and academia. For example, the celebrated Gartner hype cycle has incorporated edge intelligence as an emerging technology that will reach a plateau of productivity in the following 5–10 years [Gartner, 2018]. Major enterprises, including Google, Microsoft, Intel, and IBM, have put forth pilot projects to demonstrate the advantages of edge computing in paving the last mile of AI. These efforts have boosted a wide spectrum of AI applications, spanning from live video analytics [Ananthanarayanan et al., 2017], cognitive assistance [Ha et al., 2014] to precision agriculture, smart home [Jie et al., 2017b], and Industrial Internet of Things (IIoT) [Li et al., 2018d].

Notably, research and practice on this emerging inter-disciplinary paradigm, namely edge intelligence, is still in a very early stage. There is, in general, a lack of venue dedicated for summarizing, discussing, and disseminating the recent advances of edge intelligence, in both industry and academia. To fill this void, in this book we conduct a comprehensive and detailed survey of recent research efforts as well as our own research results on edge intelligence. As summarized in Figure 1.2, this monograph is organized into eight chapters. Specifically, we first present an overview on AI and edge computing, and then discuss the motivation, definition, and rating of edge intelligence. Next, we will review and taxonomically summarize the emerging computing architectures and enable technologies for edge intelligence model training in Chapter 2, and showcase our own research results on edge intelligence model training in Chapters 2–5. We then review and taxonomically summarize the emerging computing architectures and enable technologies for edge intelligence model inference in Chapter 6, and showcase our own research result in this branch in Chapter 7. Finally, in Chapter 8, we will discuss applications, marketplace, and open research challenges and opportunities on edge intelligence. For this book, we hope it can elicit escalating attentions, stimulate fruitful discussions, and inspire further research ideas on edge intelligence.

1.1   ARTIFICIAL INTELLIGENCE

While AI has recently ascended to the spotlight and gained tremendous attention, it is a term that was coined in 1956. Simply put, AI is centered around building intelligent machines capable of carrying out tasks as humans do. This is obviously a very broad definition, and it can encompass a wide variety of things ranging from Apple Siri to Google AlphaGo, as well as too powerful technologies yet to be invented. In simulating human intelligence, AI systems typically demonstrate at least some of the following behaviors associated with human intelligence:

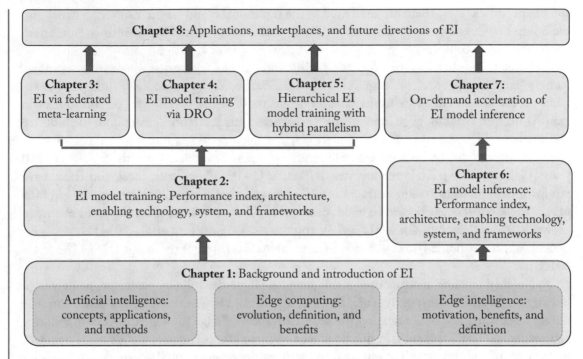

Figure 1.2: Organization of this book.

planning, learning, reasoning, problem-solving, knowledge representation, perception, motion, manipulation, and to a lesser extent social intelligence and creativity. During the past 60 years of development, AI has experienced rise, fall, and again rise and fall. The latest rise of AI after the 2010s was partially due to the breakthroughs made by deep learning, a method that has achieved human-level accuracy in some interesting areas.

1.1.1   DEEP LEARNING AND DEEP NEURAL NETWORKS

Machine learning (ML) is an effective method to achieve the goal of AI. The relationship between deep learning and AI is illustrated in Figure 1.3.

Many machine learning methodologies, as exemplified by decision tree, K-means clustering, Bayesian networks, etc., have been developed to train the machine to make classifications and predictions, based on the data obtained from the real world. Among the existing machine learning methods, deep learning, by leveraging artificial neural networks (ANN) [Svozil et al., 1997] to learn the deep representation of the data, has resulted in amazing performance in multiple applications, including image classification, face recognition, etc. Since the ANN adopted by deep learning models typically consists of a series of layers, the model is called deep neural

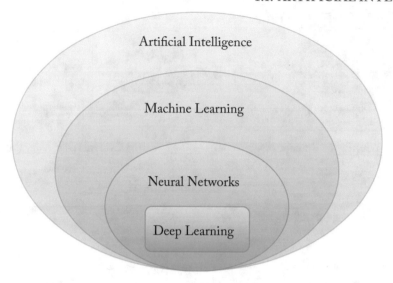

Figure 1.3: DNNs in the context of AI.

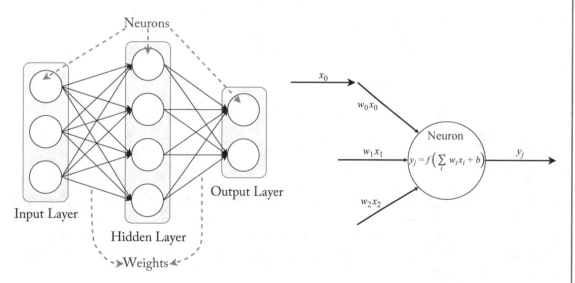

Figure 1.4: A standard composition of DL model.

network (DNN). As shown in Figure 1.4, each layer of a DNN is composed of neurons that are able to generate the nonlinear outputs based on the data from the input of the neuron.

The neurons in the input layer receive the data and "propagate" them to the middle layer (a.k.a. the hidden layer). Then the neurons in the middle layer generate the weighted sums of the input data and output the weighted sums using the specific activation functions (e.g., tanh), and

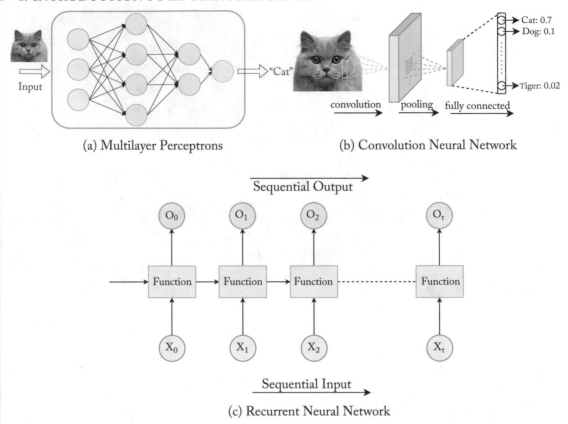

Figure 1.5: Three typical structures of DL models.

the outputs are propagated further to the output layer. The final results are presented at the output layer. With complex and abstract layers than most archetypal models, DNNs are able to learn the high-level features, enabling high precision inference in tasks. Figure 1.5 presents three popular structures of DNNs: Multilayer Perceptrons (MLPs), Convolution Neural Network (CNN), and Recurrent Neural Network (RNN).

MLPs models are the most basic DNN, which is composed of a series of fully connected layers [Collobert et al., 2011]. Different from fully connected layers in MLPs, in CNN models, the convolution layers extract the simple features from input by executing convolution operations. Applying various convolutional filters, CNN models can capture the high-level representation of the input data, making it the most popular venue for computer vision tasks, e.g., image classification (e.g., AlexNet [Krizhevsky et al., 2012], VGG network [Simonyan and Zisserman, 2014], ResNet [He et al., 2016], MobileNet [Howard et al., 2017]), and object detection (e.g., Fast R-CNN [Mao et al., 2018a], YOLO [Redmon et al., 2016], SSD [Liu et al., 2016]). RNN models are another type of DNNs, which use sequential data feeding. Shown in Figure 1.5c, the basic unit of RNN is called cell; and further, each cell consists of layers and a series of cells enables the sequential processing of RNN models. RNN models are widely used in the task of natural language processing, e.g., language modeling, machine translation, question answering, and document classification.

Deep learning represents the state-of-the-art AI technology as well as a highly resource-demanding workload that is challenging for edge computing, however. We will focus on the interaction between deep learning and edge computing. We believe that the techniques discussed can also have meaningful implications for other AI models and methods. For instance, stochastic gradient descent is a popular training method for many AI/ML algorithms (e.g., k-means, support vector machine, lasso regression) [Bottou, 2010], and the optimization techniques based on stochastic gradient descent (SGD) training introduced in a later part of the book can be also deployed for model training in other AI methods.

1.1.2 FROM DEEP LEARNING TO MODEL TRAINING AND INFERENCE

Each neuron in a DNN layer has a vector of weights associated with the input data size of the layer. Needless to say, the weights in a deep learning model need to be optimized through a training process.

In a training process for a deep learning model, the values of weights in the model are often randomly assigned initially. Then the output of the last layer represents the task result, and a loss function is set to evaluate the correctness of the results by calculating the error rate (e.g., root mean squared error) between the results and the true label. To adjust the weights of each neuron in the model, an optimization algorithm, such as SGD [Bottou, 2010], is used and the gradient of the loss function is calculated. Leveraging the back propagation mechanism [Chauvin and Rumelhart, 2013, Rumelhart et al., 1986], the error rate is propagated back across the whole neural network and the weights are updated based on the gradient and the learning rate. By feeding a large number of training samples and repeating this process until the error rate is below a predefined threshold, a deep learning model with high precision is obtained.

DNN model inference happens after training. For instance, for an image classification task, with the feeding of a large amount of training samples, the DNN is trained to learn how to recognize an image, and then inference takes real-world images as inputs and quickly draws the predictions/classifications of them. The training procedure consists of the feed-forward process and the backpropagation process. Note that the inference involves the feed-forward process only, i.e., the input from the real world is passed through the whole neural network and the model outputs the prediction. Typically, the training process requires significant computational resources for iterative parameter updates and thus is often performed in the cloud data center. The inference process, on the other hand, is executed either at the local devices (e.g., mobile phones) or in the cloud data center. Due to the resource constraint on mobile devices, many current applications (e.g., Apple Siri, Microsoft Cortana) carry out the training in the cloud data center.

1.1.3   DEEP LEARNING APPLICATIONS

Deep learning has been demonstrated to be effective in a wide spectrum of applications, including the following:

- **Computer Vision:** Given a series of images or video from the real world, the AI system learns to automatically extract the features of these inputs to complete a specific task, e.g., image classification, face authentication, and image semantic segmentation.

- **Natural Language Processing:** In this field, the task of AI is to build the system that can understand and comprehend natural language spoken by humans, e.g., natural language modeling, word embedding, and machine translation.

- **Speech Recognition:** AI system accepts the human voice as input, thereby hearing and comprehending the language in terms of sentences and meanings. It also can handle the accents from different regions, noise in background, and trendy words.

- **Intelligent Robots:** There are multiple sensors on a robot, e.g., light, heat, temperature, and sound. An AI system aggregates the sensing inputs and learns to guide the action of the robots in a real-world environment.

1.1.4   POPULAR DEEP LEARNING MODELS

For a better understanding of the deep learning and their applications, in this subsection we give an overview of various popular deep learning models.

Convolution Neural Network (CNN): For image classification, as the first CNN to win the ImageNet Challenge in 2012, AlexNet [Krizhevsky et al., 2012] consists of five convolution layers and three fully connected layers. AlexNet requires 61 million weights and 724 million MACs (Multiply-Add Computation) to classify the image with a size of 227*227. To achieve higher accuracy, VGG-16 [Simonyan and Zisserman, 2014] is trained to a deeper structure of 16 layers consisting of 13 convolution layers and 3 fully connected layers, requiring 138 million weights and 15.5G MACs to classify the image with a size of 224*224. To improve accuracy while reducing the computation of DNN inference, GoogleNet [Szegedy et al., 2015] introduces an inception module composed of different sized filters. GoogleNet achieves a better accuracy performance than VGG-16, while only requiring 7 million weights and 1.43G MACs to process the image with the same size. ResNet [He et al., 2016], the state-of-the-art effort, uses the "shortcut" structure to reach a human-level accuracy with a top-5 error rate below 5%. The "shortcut" module is used to solve the gradient vanishing problem during the training process, making it possible to train a DNN model with deeper structures. CNNs are often employed in

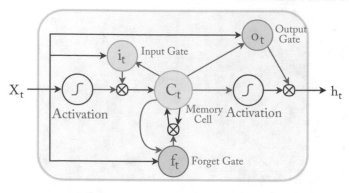

Figure 1.6: Structure of an LSTM memory cell.

computer vision. Given a series of images or video from the real world, with the utilization of CNN, the AI system learns to automatically extract the features of these inputs to complete a specific task, e.g., image classification, face authentication, and image semantic segmentation.

Recurrent Neural Network (RNN): For sequential input data, recurrent neural networks (RNNs) have been developed to address the time-series problem. The input of RNN consists of the current input and the previous samples. Each neuron in an RNN owns an internal memory that keeps the information of the computation from the previous samples. The training of RNN is based on Backpropagation Through Time (BPTT) [Werbos, 1990]. Long Short Term Memory (LSTM) [Hochreiter and Schmidhuber, 1997] is an extended version of RNNs. In LSTM, the gate is used to represent the basic unit of a neuron. As shown in Figure 1.6, each neuron in LSTM is called memory cell and includes a multiplicative forget gate, input gate, and output gate. These gates are used to control the access to memory cells and to prevent them from perturbation by irrelevant inputs. Information is added or removed through the gate to the memory cell. Gates are different neural networks that determine what information is allowed on the memory cell. The forget gate can learn what information is kept or forgotten during training. RNNs have been widely used in natural language processing due to the superiority of processing the data with an input length that is not fixed. The task of the AI here is to build a system that can comprehend natural language spoken by humans, e.g., natural language modeling, word embedding, and machine translation.

Generative Adversarial Network (GAN): As illustrated in Figure 1.7, generative adversarial networks (GANs) [Goodfellow et al., 2014a] consist of two main components, namely the generative and discriminator network (i.e., generator and discriminator). The generator is responsible for generating new data after it learns the data distribution from a training dataset of real data. The discriminator is in charge of classifying the real data from the fake data generated by the generator. GAN is often deployed in image generation, image transformation, image synthesis, image super-resolution, and other applications. The target of GANs is that the generator

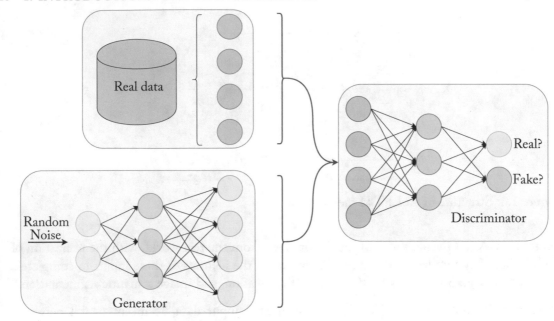

Figure 1.7: Composition of a adversarial network.

is optimized to generate input data that is deceiving the discriminator, i.e., the discriminator cannot recognize the data whether it is fake or real.

Deep Reinforcement Learning (DRL): Deep Reinforcement Learning (DRL) is composed of DNNs and reinforcement learning (RL) (Figure 1.8). The goal of DRL is to create an intelligent agent that can find efficient policies to maximize the rewards of long-term tasks with controllable actions. DRL has been used to solve various scheduling problems, such as decision problems in games, rate selection of video transmissions, etc.

In the DRL approach, the RL searches for the optimal policy of actions over states from the environment, and the DNN is in charge of representing a large number of states and approximating the action values to estimate the quality of the action in the given states. The reward is a function to represent the distance between the predefined requirement and the performance of an action. Through continuous learning, the agent of DRL model can be used for various tasks, e.g., gaming [Mnih et al., 2015].

1.2   EDGE COMPUTING

In this section, we first review the development history of edge computing and present a brief chronology of the major advances in its evolution. Then, we discuss various benefits or edge computing.

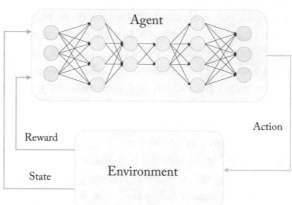

Figure 1.8: Sketch of a deep reinforcement learning model.

1.2.1   EVOLUTION OF EDGE COMPUTING

There is a general consensus that edge computing can be traced back to the late 1990s, when MIT researchers introduced the content delivery network (CDN) [Dilley et al., 2002], and founded Akamai to commercialize this technology to mitigate Internet congestion and accelerate web performance. A CDN refers to a cluster of proxy server nodes in close proximity to endusers of the Internet. By prefetching web contents from remote datacenters and caching them locally, CDN is able to deliver high availability, high performance, and content customization (e.g., location-based advertising). Since CDN enables substantial bandwidth savings for content providers, it serves a large portion of the Internet content today, ranging from web objects (text, graphics and scripts) and downloadable objects (media files, software, documents) to streaming media. Around the same time period, researchers in CMU demonstrated the potential of computation offloading for mobile computing. Specifically, by offloading the computation of a speech recognition application to a remote server, they showed how to reduce latency and improve battery life of resource-limited mobile devices [Flinn and Satyanarayanan, 1999]. Later in 2001, they generalized these techniques to the concept of "cyber foraging" [Satyanarayanan et al., 2001], for the amplification of exploiting wired hardware infrastructures (referred to as surrogates, e.g., desktop computers) to dynamically augment wireless mobile device's computing capabilities, via short-range wireless peer-to-peer communication technology.

In spite of the significance of cyber foraging on improving the performance and energy-efficiency of mobile devices, there were also many practical technical barriers. For example, who would provide the infrastructures for cyber foraging? Where to deploy the infrastructures and how to discover them? How to design the business model for cyber foraging? The emergence of cloud computing in the mid-2000s offers a great answer to these open questions. In particular, in the form of Infrastructure-as-a-Service, major companies, such as Amazon, Microsoft, and Google, package computing resources in their mega-scale datacenters into virtual machines that can be ubiquitous accessible to users who have connections to the Internet in an on-demand manner. This has made the cloud the most appealing infrastructure to be leveraged by mobile devices to argument their resource capacities. With cloud computing, many mobile applications, as exemplified by Apple Siri and Google Assistant, offloading the computation intensive speech-recognition workload to the cloud for processing, have opened up a research field known as mobile cloud computing [Dinh et al., 2013].

Many emerging IoT applications, such as autonomous driving and augmented reality, need intensive computation to accomplish object tracking, content analytics and intelligent decision in a real-time manner, in order to meet the requirements for safety, accuracy, performance and user experience. That is to say, *the necessity of real-time edge intelligence for these IoT applications dictates that decision making should take place right here right now at the network edge.* If enormous amounts of data had to be forwarded on up the chain to more capable servers that typically live in the cloud for advanced data analysis, the requirements in terms of high bandwidth and low latency (often millisecond end-to-end latency) would be extremely demanding and stringent. Needless to say, cloud computing can utilize abundant computing resources for handling complex tasks, but one significant challenge therein is to meet the stringent low latency requirement, due to unpredictable network delay in transporting time-sensitive data traffic through the Internet. The utility of mobile cloud computing had been evidenced by some empirical studies conducted by CMU. Specifically, for a highly interactive visualization application Quake-Viz that requires high output frame rate, it is observed that local machine with hardware graphics acceleration brings better smoothness than a remote compute server [Lagar-Cavilla et al., 2007]. Independently, another empirical study conducted in 2006 also showed that the usability of a highly interactive photo editing application suffers unacceptably even at moderate WAN latency (100 ms round-trip time) and very good bandwidth (100 Mbps) [Tolia et al., 2006]. In conclusion, *it is expected that a high percentage of IoT-created data will have to be stored, processed and analyzed upon close to, or at the edge of the network.*

To tackle the challenging latency issue in mobile cloud computing, in October 2008, Microsoft Researchers, together with researchers from CMU, AT&T, Intel, and Lancaster University, came up with a new computing paradigm called "cloudlet." Different from the cloud which is in the remote network core, a cloudlet can be accessed by nearby mobile devices via low-latency, one-hop, high-bandwidth wireless networks. Thanks to physical proximity, cloudlets can enable low end-to-end response time for the offloaded mobile applications [Satyanarayanan et al., 2009]. Later in 2009, they published this cloudlet concept in the paper "The Case for VM-Based Cloudlets in Mobile Computing," which has received more than 2500 citations by the end of 2018. Subsequent works on empirical measurements and prototype implementation for latency-sensitive and compute-intensive applications, conducted in the early 2010s [Cuervo et al., 2010, Ha et al., 2014], have demonstrated the advantages of cloudlets.

Following the concept of cloudlet, in 2012, Cisco proposed the concept of fog computing [Bonomi et al., 2012], which refers to the continuum computing infrastructure between end

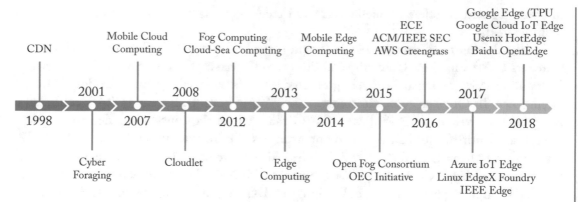

Figure 1.9: The timeline of the evolution of edge computing.

devices and traditional cloud-scale datacenters. Unlike cloudlet which was driven by mobile applications, the primitive of fog computing was IoT applications, e.g., connected vehicles, smart grids, and smart cities. In 2010, Institute of Computing Technology (ICT) of Chinese Academy of Science (CAS) initialed the cloud-sea computing systems project [Xu, 2014], aiming to augment conventional cloud computing by cooperation and integration of the cloud-side systems and the sea-side systems, where the "sea-side" refers to an augmented client side consisting of human facing and physical world facing devices and subsystems. In 2013, Pacific Northwest National Laboratory (PNLL) proposed the term edge computing for the first time. In the 2-page report [Lomathe, 2012], the motivation and main research areas of edge computing were briefly introduced.

Since 2014, edge computing has stepped into a fast track of development and promotion, as sketched in Figure 1.9. In late 2014, the European Telecommunications Standards Institute (ETSI) created a new Industry Specification Group (ISG) to develop technical standards to tackle the complex questions associated with mobile edge computing, attracting the participation of more than 30 IT companies. In the following year, the Open Edge Computing initiative was launched in by Vodafone, Intel, and Huawei in partnership with CMU and expanded a year later to include more industrial members. Also in 2015, the Open Fog Consortium was created by Cisco, Microsoft, Intel, Dell, and ARM in partnership with Princeton University, and has since expanded to include many other companies. In 2016, Edge Computing Consortium (ECE) was formed in China, with the position of serving as an industrial cooperative platform to promote open cooperation in the operation technology (OT) and information-communication-technology (ICT) fields, nurture the industry's best application practices, and advance sound and sustainable development of the edge computing industry. By the end of 2018, more than 200 IT companies around the world joined ECE. In early 2019, 18 vendors and organizations have signed a cooperation agreement to form the Edge Computing Consortium European, aiming to create a standard reference architecture and technology stack that can be deployed across multiple fields.

The driving force from academia should never be neglected. In Summer 2015, two researchers at Arizona State University and Princeton University founded Smartiply Inc., a edge Computing startup company delivering boosted network connectivity and embedded artificial intelligence. In October 2016, researchers from Wayne State University (WSU) published a comprehensive vision paper [Shi et al., 2016] that defined the concept, verified the applications, and identified the challenges and opportunities of edge computing. By the end of 2018, this paper had been cited for nearly 700 times. Also in October 2016, the First ACM/IEEE Symposium on Edge Computing (SEC) together with NSF Workshop on Grand Challenges in Edge Computing were co-held in Washington, DC. In June 2017, the First IEEE International Conference on Edge Computing (EDGE) was held in Honolulu, HI. In July 2018, the first USENIX Workshop on Hot Topics in Edge Computing (HotEdge) was held in Boston, MA. Since 2017, workshops on edge computing begun to be held in-conjunction with influential conferences as exemplified by ACM SIGCOMM and IEEE INFOCOM.

Realizing the great business opportunities provided by edge computing, a crowd of companies have flooded into the edge computing market. In December 2016, Amazon made foray into edge computing with AWS Greengrass, a service that brings local computing, messaging, data caching, sync, and ML inference capabilities to edge devices. In April 2018, Microsoft announced that it would invest $5 billion in IoT and edge computing technologies over the next four years. Google has also recently released the edge TPU AI chips and Cloud IoT Edge platform to bolster its edge computing strategy. For the internet giant Baidu from China, it also launched and open-sourced it edge computing platform OpenEdge in 2018. A more comprehensive and holistic profile of the marketplace of edge intelligence will be given in Chapter 8. In a nutshell, the game change on edge computing advocated by the giant players in the IT sector indicates that edge computing is stepping into a booming era.

1.2.2   BENEFITS OF EDGE COMPUTING

Even though the notion of "edge computing" was first introduced in 2013, there is still a lack of a universally acceptable definition for edge computing. Specifically, various organizations have defined edge computing from different perspectives.

In 2013, when edge computing was first introduced in a two-page report [Lomathe, 2012], Pacific Northwest National Laboratory (PNLL) gave the definition that "Edge computing is pushing the frontier of computing applications, data, and services away from centralized nodes to the logical extremes of a network. It enables analytics and knowledge generation to occur at the source of the data. This approach requires leveraging resources that may not be continuously connected to a network such as laptops, smartphones, tablets and sensors." This definition focuses on the on-premise capability that enables service continuity even without network connection to the cloud. Since at that time, connectivity was not ubiquitously or economically available.

In 2015, in an introductory technical white paper [Hu et al., 2015] released by ETSI, mobile edge computing is defined as "a key technology toward 5G, which provides an IT service environment and cloud-computing capabilities at the edge of the mobile network, within the Radio Access Network (RAN) and in close proximity to mobile subscribers. The aim is to reduce latency, ensure highly efficient network operation and service delivery, and offer an improved user experience." As a standard organization for telecommunications, ETSI defines edge computing as a technical enabler of the upcoming 5G which requires low latency, and stresses the superiority of edge computing on reducing the latency, and thus improving mobile experience and network operation. Moreover, for the first time, EISI pointed out that the edge node should be placed within the RAN to ensure the physical proximity.

In 2016, in the positioning paper [Shi et al., 2016], edge computing is defined as "the enabling technologies allowing computation to be performed at the edge of the network, on downstream data on behalf of cloud services and upstream data on behalf of IoT services. Here 'edge' as any computing and network resources along the path between data sources and cloud data centers." Clearly, this definition from WSU is highly related to IoT, since the latter has emerged as a top-scenario at that time. With the push of data from cloud services and pull of data from IoT, the edge of the network is changing from data consumer to data producer as well as data consumer. The definition above highlights the potential of edge computing on addressing the concerns of response time requirement, battery life constraint, bandwidth cost saving, as well as data safety and privacy.

As a cloud computing service provider, in 2018, Microsoft-defined edge computing as [mse] "where compute resources, ranging from credit-card-size computers to micro data centers, are placed closer to information-generation sources, to reduce network latency and bandwidth usage generally associated with cloud computing. Edge computing ensures continuation of service and operation despite intermittent cloud connections." By this definition, edge computing is the paradigm that the information-generation source makes use of the available and widely dispersed computing resources for computation, and thus to reduce the network latency and bandwidth usage. The edge node can be varying in size: as smaller as a credit-card-size computer, and as larger as a micro datacenter with several server racks.

While the detailed definition of edge computing varies from different organizations, physical proximity is the most essential characteristic emphasized by those definitions. To ensure physical proximity, an edge node can be a server attached to an access point (e.g., WiFi, router, cellular tower), a network gateway, or a micro-datacenter in practice. Besides, D2D (device-to-device) offloading [Asadi et al., 2014] that offloads data or tasks among nearby end-devices also ensures physical proximity and has the virtues of high data rate, energy-efficiency. Therefore, D2D offloading is also generally recognized as a form of edge computing [Chen et al., 2017a, Mao et al., 2017c, Wang et al., 2017b]. Notably, the physical proximity between the computing and information-generation sources have several benefits compared to the traditional cloud-based computing paradigm, including low-latency, energy-efficiency, privacy protection, reduced bandwidth consumption, on-premises, and location awareness. We next elaborate these benefits as follows.

Low Latency: Humans are naturally sensitive to delays in the critical path of interaction. For a mobile service, the end-to-end latency perceived by the end-users typically consists of four components: data transmission delay, network propagation delay, queuing delay, and computing delay [Kurose and Ross, 2009]. By pushing the computing resources from the remote network core to the network edges that are in very close proximity to the end users, the network propagation delay can be significantly reduced, due to the greatly shortened network distance. Besides, with cloud computing, the information flow is required to pass through multiple network gateways, routers and switches distributed at several networks including the access network, backhaul network, and WAN, where traffic control, routing, and other network-management operations can contribute to excessive queuing delay. Moreover, in edge computing, the end users can access the edge computing node via metropolitan-area-network (MAN) or local-area-network (LAN) whose bandwidth is typically sufficient and stable. In a sharp contrast, in cloud computing, the information should traverse the WAN whose bandwidth are scares, limited, and unstable (in terms of jitters) [Jalaparti et al., 2016]. This results in much shorter data transmission delay in edge computing than that in cloud computing. Therefore, for mobile applications that do not require excessive amount of computing resources, edge computing can significantly reduce the overall end-to-end latency. Extensive recent research efforts have empirically demonstrated the low-latency of edge computing for various mobile applications. For example, for a mobile augmented reality application, CMU showed that cloud computing typically takes 250–1000 ms, while edge computing can finish the tasks within 140 ms [Ha et al., 2013]. For a mobile face recognition application, researchers from The College of William and Mary observed that edge computing is able to reduce the response time from 900 ms to 169 ms [Yi et al., 2015]. Undoubtedly, the low latency of edge computing makes it appealing to emerging mission-critical applications as exemplified by augmented reality, interactive game, IIoT, and Internet of Vehicles (IoV).

Energy Efficiency: For mobile and embedded devices, the energy consumption is typically dominated by CPU and network interface that performs computing and data communication, respectively. Due to the limited size of mobile and embedded devices, due to the constraint on the physical size, the capacity of the battery is limited. Therefore, energy efficient task execution is of strategical importance for mobile and embedded devices. Compared to on-device computing, edge computing eliminates the computing energy consumption of the devices. When compared to the cloud computing, edge computing significantly reduces the energy consumption of the network interface incurred by the data transmission. Similarly, many empirical measurements have been conducted to compare the energy-efficiency of edge comput-

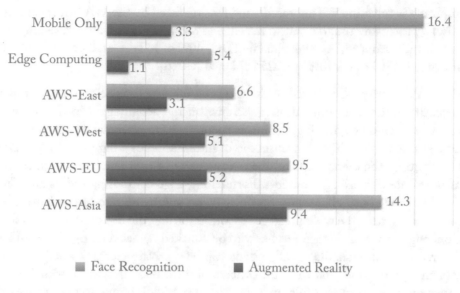

Figure 1.10: The per-operation energy consumption (measured by Jouls) of a face recognition and an argumented reality application on a mobile device, in which an image from the device is transmitted over a Wi-Fi first hop to an edge node or Amazon Web Services (AWS) datacenters at various regions.

ing and cloud computing. For example, for an augmented reality and a mobile face recognition application, researchers in CMU measured the energy consumption of on-device computing, edge computing, and cloud computing with datacenters in various regions [Ha et al., 2013]. The results are plotted in Figure 1.10, from which we observe that edge computing has better energy efficiency than on-device computing and cloud computing.

Bandwidth Efficiency: Over the past decade, we have witnessed a wide range of ubiquitous IoT devices as exemplified as surveillance cameras, smart sensors, wearables, and IoVs connected to the Internet. According to forecast of Cisco Visual Networking Index (2017–2022) [Cisco, 2017], there are an estimated number 18 billion networked devices across the global in 2017, up 22% per year, and will reach 28.5 billions by 2022. As a concrete example, the emerging wearable devices have seen a 44.4% increase in shipments for 2016, compared to the 80 million device shipments in 2015, and the yearly shipments are expected to reach 200 million in 2019 [Seneviratne et al., 2017]. As a result of the booming connectivity, the data volume will grow explosively. By 2020, data generated by the ubiquitous IoT devices worldwide is expected to reach 500 ZB, meaning that each person will produce 1.7 MB data per second. While in a sharp contrast, the pace of the development of WAN infrastructure that is traditional used to ship the IoT data to the cloud has slowed down [Viswanathan et al., 2016, Vulimiri et al., 2015].

Edge computing is releasing the unprecedented tension faced by the WAN infrastructure now. Specifically, by processing the IoT data at the edge of the WAN, the traffic traversing the WAN can be significantly reduced. As forecasted by IDC, 40% of IoT-created data will be processed at network edge by 2018 and 50% by 2025 [Shi et al., 2016].

Privacy Protection: Compared to cloud computing, the capability of enhancing the privacy and security of the user information is also an attractive advantage owned by edge computing. Specifically, in cloud computing, the infrastructure, i.e., mega-scale datacenters are susceptible to attacks due to their high concentration of information of a big pile of users. Moreover, in cloud computing, the ownership and the management of user data are separated, incurring the threat of private data leakage and loss. Fortunately, edge computing brings an effective solution to address the above challenges. Specifically, on one hand, due to the large-in-number but small-in-scale nature, the widespread geo-distribution and the less concentration of valuable information, edge servers are much less likely to be attacked by hackers. Besides, while a public cloud is shared by multi-tenants, an edge node can be a private cluster owned by a factory, a company or an organization. This eases the concern of the leakage of user information, since the administrator of the edge cluster fully manages the authorization and access control of service requests, without the need of an external party such as the public cloud.

Context Awareness: By placing the computing resources and services at the edge of the network, the operator of the edge computing services can leverage low-level signaling information to determine real-time information such as user behaviors, locations, and environments [Perera et al., 2015]. This context information allows the delivery of context-aware services to end users, ranging from location-based services to local points-of-interest businesses and network QoS optimization, etc. For example, a retail shopping AR application can benefit not only from knowing the relative location of a user (e.g., the appliance department), but also from knowing that the user is interested in a specific type of appliance (e.g., a microwave oven). Similarly, a video streaming application might optimize its content caching strategy by knowing where the user is and knowing whether a person interacting with the user is someone the user interacted with before [Lobo et al., 2014, Wang et al., 2014].

On Premise: As a local computing unit, edge servers can run isolated from the rest of the network, while having access to local resources. This becomes particularly important for failure management and defense. If a cloud service becomes unavailable due to network failures, datacenters failures (e.g., power outages) or denial-of-service attacks, the fallback service on the edge can temporarily mask the failure, maintaining the normal operation the services and data. Clearly, this advantage is quite flavored by mission-critical applications (e.g., mobile payment) that suffer great loss due to temporal failures.

1.3   EDGE INTELLIGENCE

The marriage of edge computing and artificial intelligence has given rise to the birth of edge intelligence. Broadly speaking, **edge intelligence (EI) refers to the model training or inference at the network edge by exploiting available data and computing resources across the hierarchy ranging from end devices to edge nodes to cloud datacenters.** In this section, we discuss the motivation, benefits, and definition of edge intelligence.

1.3.1   MOTIVATION AND BENEFITS OF EDGE INTELLIGENCE

The fusion of AI and edge computing is natural, since there are strong intersections between edge computing and AI. Specifically, edge computing aims at coordinating a multitude of collaborative edge devices and servers to process the generated data in proximity; and AI strives for simulating intelligent human behavior in devices/machines by learning from data. Besides enjoying the benefits of edge computing (see Section 1.2.2), pushing AI to the edge further benefits each other in the following aspects.

Our first observation is that data generated at the network edge need AI to fully unlock their potential. In recent years, we have witnessed a skyrocketing number and types of end devices ranging from surveillance cameras, smart sensors, and wearables to IoVs being connected to the Internet. As a result of the proliferation of these diverse devices, large volumes of multi-modal data (e.g., audio, picture and video) of physical surroundings are continuously sensed at the device end. In this context, AI will be functionally necessary due to its ability to quickly analyze those huge data volumes and extract insights from them for high-quality decision making. As one of the most popular AI techniques, deep learning brings the ability to automatically identify patterns and detect anomalies in the data sensed by the edge device, as exemplified by population distribution, traffic flow, humidity, temperature, pressure and air quality. The insights extracted from the sensed data are then fed to the real-time predictive decision-making (e.g., public transportation planning, traffic control and driving alert) in response to the fast changing environments, increasing the operational efficiency. Compared to traditional intelligence approaches based on the monitoring of numeric thresholds to be crossed, deep learning approaches improve the optimal decision making with greater accuracy. Inspired by the efficacy, predictive AI capability has been integrated with major general-purpose and industrial IoT platforms, such as Microsoft Azure IoT, IBM Watson IoT, Amazon AWS IoT, GE Predix, and PTC ThingWorx. As forecasted by Gartner, more than 80% of enterprise IoT projects will include an AI component by 2022, up from only 10% today.

On the flip side of the coin, edge computing is able to enrich AI with big data and application scenarios. It is widely recognized that the driving force behind the recent booming of deep learning is four-folds: algorithm, hardware, data, and application scenarios. While the effect of algorithm and hardware on the development of deep learning is intuitive, the roles of data and application scenarios have been mostly overlooked. Specifically, to improve the performance of deep learning algorithm, the most commonly adopted approach is to refine the DNN with more layers of neurons. By doing this, we need to learn more parameters in the DNN, and so does the data required for training increase. By reviewing the timing of the most publicized AI breakthroughs over the past 30 years, it is observed that the average elapsed time between key algorithms and corresponding advances was about 18 years, whereas the average elapsed time between key datasets and corresponding advances was less than 3 years [Quanto, 2016]. This definitely demonstrates the importance of data on the development of AI. Having recognized the importance of data, the next problem is, where is the data from. Traditionally, data is mostly born and stored in the mega-scale datacenters. Nevertheless, with the rapid development of IoT, the trend is reversing now. Cisco GCI estimates that nearly 850 ZB will be generated by all people, machines, and things by 2021, up from 220 ZB generated in 2016 [Cisco, 2016]. In contrast, the global datacenter traffic will only reach 20.6 ZB by 2021, up from 6.8 ZB in 2016. Clearly, in the following years, IoT would continue to fuel the booming of AI.

While edge computing and AI complement each other from a technical perspective, their applications are also mutually beneficial.

On one hand, AI democratization requires edge computing as a key infrastructure. AI technologies have witnessed great success in many digital products or services in our daily life, e.g., online shopping, service recommendation, video surveillance, smart home devices, etc. AI is also a key driving force behind emerging innovative frontiers, such as self-driving cars, intelligent finance, cancer diagnosis, and medicine discovery. Beyond the above examples, to enable a richer set of applications and push the boundaries of what's possible, AI democratization or ubiquitous AI [Microsoft, 2016] has been declared by major IT companies, with the vision of "making AI for every person and every organization at everywhere." To this end, AI should go "closer" to the people, data and end devices. Clearly, edge computing is more competent than cloud computing in achieving this goal. First, compared to the cloud datacenter, edge servers are in closer proximity to people, data source and devices. Second, compared to cloud computing, edge computing is also more affordable and accessible. Finally, edge computing has potential to provide more diverse application scenarios of AI than cloud computing. Due to these advantages, edge computing is naturally a key enabler for ubiquitous AI.

On the other hand, edge computing can be further popularized with AI applications. During the early development of edge computing, there has always been the concern in the cloud computing community with which high-demand applications edge computing could take to the next level that cloud computing could not, and what are the killer applications of edge computing. To clear up the doubt, the research group in Microsoft, who co-introduced the concept of cloudlet [Satyanarayanan et al., 2009], has conducted continuous exploration on what kinds should be moved from the cloud to the edge since 2009, ranging from voice command recognition, AR/VR and interactive cloud gaming to real-time video analytics. By comparison, real-time video analytics is envisioned to be a killer application for edge computing [Ananthanarayanan et al., 2017]. As an emerging application built on top of computer vision, real-time video analytics continuously pulls high-definition videos from surveillance cameras, and requires high computation, high bandwidth, high privacy, and low-latency to analyze the videos. Clearly, the viable approach that can meet these strict requirements is edge computing. Interdependently, in CMU, to popularize the concept of cloudlet, researchers have mainly focused on a dozen of cognitive assistance applications that bring AI technologies such as computer vision, speech recognition, natural language processing to the inner loop of human cognition and interaction. Looking back to the above evolution of edge computing, it can be easily seen that AI applications have played a crucial role in the popularization of edge computing. This is due to the fact that many mobile and IoT-related AI applications represent a family of practical applications that are computation- and energy-intensive, privacy- and delay-sensitive, and thus naturally align well with edge computing.

Due to the superiority and necessity of running AI application on the edge, edge AI has recently received great attention. In December 2017, in a white paper "A Berkeley View of Systems Challenges for AI" [Stoica et al., 2017] published by UC Berkeley, the cloud-edge AI system is envisioned as an important research direction to achieve the goal of mission-critical and personalized AI. In August 2018, edge AI emerges in the Gartner Hype Cycle for the first time. According to Gartner's prediction, edge AI is still in the innovation trigger phase, and it will reach a plateau of productivity in the following 5–10 years. In the industry, many pilot projects have also been carried out toward edge AI. Specifically, on the edge AI service platform, the traditional cloud providers, such as Google, Amazon, and Microsoft, have launched service platforms to bring the intelligence of Google Cloud to the edge, by enabling end devices to run ML inferences with pre-trained models locally. On edge AI chips, various high-end chips designated for running ML models have been make commercially available on the market, as exemplified by Google Edge TPU, Intel Nervana NNP, Huawei Ascend 910, and Ascend 310.

1.3.2   SCOPE AND RATING OF EDGE INTELLIGENCE

While "edge AI" or "edge intelligence" is still in its infancy stage, explorations and practical applications along this avenue started earlier on. As aforementioned, in 2009, to demonstrate the benefits of edge computing, Microsoft has built an edge-based prototype to support mobile voice command recognition, an AI application. Albeit the early begin of exploration, there is still not a formal definition for edge intelligence.

Currently, most organizations and presses refer to edge intelligence as the paradigm of running AI algorithms locally on an end device, with data (sensor data or signals) that are created on the device. While this represents the current most common approach (e.g., with high-end AI chips) toward edge intelligence in the real world, it is crucial to note that this definition greatly narrows down the solution scope of edge intelligence. Running computation intensive algorithms as exemplified by DNN models locally is very resource-intensive, requiring high-end processors to be equipped in the device. Clearly, such stringent requirement not only increases the cost of edge intelligence, but is also incompatible and unfriendly to existing legacy end devices that have limited computing capacities.

In this book, we submit that the scope of edge intelligence should not be restricted to running AI models solely on the edge server or device. In fact, as demonstrated by a dozen of recent studies, for DNN models, running them with edge-cloud synergy can reduce both the end-to-end latency and energy consumption, when compared to the local execution approach. Due to these practical advantages, we believe that such collaborative hierarchy should be integrated into the design of efficient edge intelligence solutions. Further, existing thoughts on edge intelligence mainly focus on the inference phase (i.e., running the AI model), assuming that the training of the AI model is performed in the power cloud datacenters, partially because the resource consumption of the training phase significantly overweights the inference phase. However, this would dictate that the enormous amount of training data should be shipped from devices or edges to the cloud, incurring prohibitive communication overhead as well as the concern on data privacy.

Taking a forward-looking perspective, we believe that edge intelligence should be the paradigm that fully exploits the available data and resources across the hierarchy of end devices, edge nodes and cloud datacenters to optimize the overall performance of training and inferencing a DNN model. This indicates that edge intelligence does not necessarily mean that the DNN model is fully trained or inferenced at the edge, but can work in a cloud-edge-device coordination manner via data offloading. Specifically, according to the amount and path length of data offloading, we rate edge intelligence into six levels, as shown in Figure 1.11. Specifically, the definition of various levels of edge intelligence is given as follows.

- Cloud Intelligence: training and inferencing the DNN model fully in the cloud.

- Level-1 – Cloud-Edge Co-Inference and Cloud Training: training the DNN model in the cloud, but inferencing the DNN model in an edge-cloud cooperation manner. Here edge-cloud cooperation means that data is partially offloaded to the cloud.

- Level-2 – In-Edge Co-Inference and Cloud Training: training the DNN model in the cloud, but inferencing the DNN model in an in-edge manner. Here in-edge means that the model inference is carried out within the network edge, which can be realized by fully or partially offloading the data to the edge nodes or nearby devices (via D2D communication).

- Level-3 – On-Device Inference and Cloud Training: training the DNN model in the cloud, but inferencing the DNN model in a fully local on-device manner. Here on-device means that no data would be offloaded.

- Level-4 – Cloud-Edge Co-Training & Inference: training and inferencing the DNN model both in the edge-cloud cooperation manner.

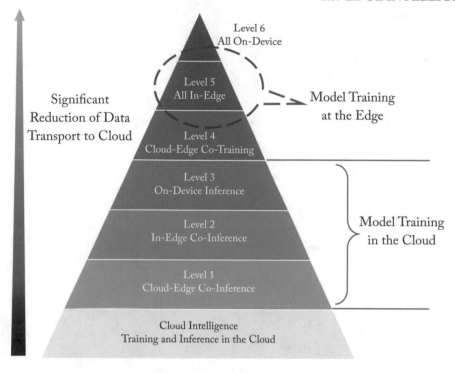

Figure 1.11: A six-level rating for edge intelligence.

- Level-5 – All In-Edge: training and inferencing the DNN model both in the in-edge manner.

- Level-6 – All On-Device: training and inferencing the DNN model both in the on-device manner.

Clearly, as the level of edge intelligence moves higher, the amount and path length of data offloading decreases. As a result, the transmission latency of data offloading decreases, the data privacy increases and the WAN bandwidth cost reduces. Nevertheless, this is achieved at the cost of increased computational latency and energy consumption. This conflict indicates that there is no "one-level-fits-all" in general; instead, the "best-level" edge intelligence is application-dependent and it should be determined by jointly considering multi-criteria such as latency, energy efficiency, privacy and WAN bandwidth cost. We believe that level-6 edge intelligence is the ultimate goal, and it is the future. In the later chapters, we will review the commercial solutions as well as enabling techniques for different levels of edge intelligence.

CHAPTER 2

Edge Intelligence via Model Training

In this chapter, we focus on distributed training of DNN at the edge, including the architectures, key performance indicators, enabling techniques, and existing systems and frameworks.

2.1 ARCHITECTURES

We clasify the architectures of distributed DNN training into three categories: Centralized, Decentralized, and Hybrid (Cloud-Edge-Device), as illustrated by subfigures (a), (b), and (c) in Figure 2.1, respectively. The cloud refers to the central datacenter whereas the end devices are represented by mobile phones, cars, and surveillance cameras, which are also data sources. For the edge server, we use base stations as the legend. As shown in Figure 2.1, all three architectures involve cooperation between different kinds of devices. However, only the devices with a neural network mark process DNN training, either partially or wholy.

2.1.1 CENTRALIZED

Figure 2.1a describes a centralized DNN training, where the DNN model is trained in the cloud datacenter. The data for training is generated and gathered from distributed end devices such as mobile phones, cars and surveillance cameras. Once the data arrives, the cloud datacenter performs DNN training using these data. Therefore, the system based on the centralized architecture can be associated with Cloud Intelligence (Level-1, Level-2, or Level-3) in Figure 1.11, according to the specific inference mode that the system employs. Under this centralized mode, it is easy to deploy the DNN model since the DNN model is only placed on the cloud datacenter. However, the centralization of training data may cost a considerable communication overhead due to the possible large data size and the unpredictable network connection. Besides, the data with sensitive personal information is gathered in the cloud datacenter, which inevitably points to privacy issues, especially when the data geo-distributes globally in various regions.

2.1.2 DECENTRALIZED

Users expect to benefit from AI-based services while keeping their own personal privacy. Unfortunately, the centralized mode fails to provide these benefits at the same time. To avoid this dilemma, the decentralized mode is introduced. As Figure 2.1b shows, there is no centralized

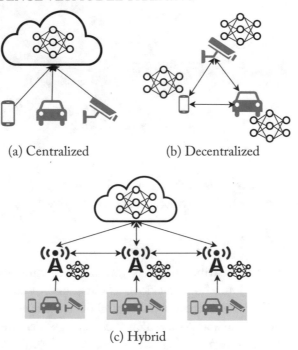

(a) Centralized   (b) Decentralized

(c) Hybrid

Figure 2.1: Three architecture modes for distributed training.

node in the decentralized network, i.e., all the computing nodes perform equal roles. It is re-markable that all the computing nodes are with a neural network mark, implying that each of them can perform DNN training. Under the decentralized mode as shown in Figure 2.1b, each computing node trains its own DNN model locally with local data, which preserves private information locally. To obtain the global DNN model by sharing local training improvement, nodes in the network communicate with each other to exchange the local model updates. In this mode, the global DNN model can be trained without the intervention of the cloud datacenter, corresponding to the Level-5 edge intelligence defined in Figure 1.11.

2.1.3   HYBRID

The hybrid mode combines the centralized mode and the decentralized mode. As shown in Figure 2.1c, as the hub of the architecture, the edge servers may train the DNN model by either decentralized updates with each other or centralized training with the cloud datacenter, thus the hybrid architecture covers Level-4 and Level-5 in Figure 1.11. The hybrid architecture is also called as Cloud-Edge-Device training due to the involved roles. The private data is only gathered in the edge servers, resulting in the privacy preservation, weaker than the decentralized architecture but stronger than the centralized architecture. Besides, the distributed training at the edge consumes less communication overhead comparing to the other two modes. Note that the hybrid architecture is not limited in the Figure 2.1c in practical deployment. It is flexible to adapt to the application scenario.

2.2   KEY PERFORMANCE INDICATORS

To better assess a distributed training method, we present six key performance indicators.

2.2.1   TRAINING LOSS

In essence, the DNN training process solves an optimization problem that seeks to minimize the training loss. Since the training loss captures the gap between the learned (e.g., predicted) value and the labeled data, it indicates how well the trained DNN model fits the training data. Therefore, it is expected that the training loss can be minimized. Training loss (and generalization error) is mainly affected by training samples and training methods.

2.2.2   CONVERGENCE

Convergence in machine learning often refers to achieving an error equal to (or close to) local/global minimum. In particular, decentralized learning algorithms need to reach consensus indicator. Intuitively, a decentralized method works well only if the distributed training processes converge to a consensus. When the decentralized approach is used for training, the convergence depends on the way the gradient is synchronized and updated.

2.2.3   PRIVACY

When training the DNN model by using the data originated at a massive of end devices, the raw data or intermediate data should be transferred out of the end devices. Obviously, it is inevitable to deal with privacy issues in this scenario. To preserve privacy, it is expected that less privacy-sensitive data is transferred out of the end-devices. Whether privacy protection is implemented depends on whether the raw data is offloaded to the edge. More specifically, in a recent work [Zhou et al., 2020] on edge intelligence model training, the level of privacy preservation is quantitatively measured by the ratio of the amount of offloaded data to the total amount of data. With such a measure, a smaller ratio indicates stronger privacy preservation.

2.2.4   COMMUNICATION COST

Training the DNN model is data-intensive, since the raw data or intermediate data should be transferred across the nodes. Intuitively, this communication overhead increases the training latency, energy, and bandwidth consumption. Communication overhead is affected by the size of the original input data, the way of transmission and the available bandwidth.

2.2.5   LATENCY

Arguably, latency is one of the most fundamental performance indicators of distributed DNN model training, since it directly influences when the trained model is available for use. The latency of the distributed training process includes the computation latency and the communication latency. The computation latency hinges heavily upon the capability of the edge nodes. The communication latency may vary from the size of transmitted raw or intermediate data, and the bandwidth of network connection.

2.2.6   ENERGY EFFICIENCY

When training the DNN model in a decentralized manner, both the computation and communication process consume enormous energy. However, for most end-devices, they are energy-constrained. As a result, it is highly desirable that the DNN model training is energy-efficient. Energy efficiency is mainly affected by the size of the target training model and resources of the used devices.

It is worth noting that the performance indicators training loss and convergence are common objectives, thus they may not be explicitly claimed by some literature on DNN training. Note that the convergence of a method is influenced by many other factors such as the training dataset, the selected optimizer and the provided computing resources. Generally, for the decentralized methods, the faster the method converges, the better the method is.

2.3   ENABLING TECHNOLOGIES

In this subsection, we review a few archetypal enabling technologies for improving one or more of the aforementioned key performance indicators when training edge intelligence model. Table 2.1 summarizes the highlights of each enabling technology.

2.3.1   FEDERATED LEARNING

Due to the unprecedented accuracy, deep learning methods have turned themselves into an essential role in AI-based services on the Internet. While enjoying their advantage, it is inevitable to meet privacy issues. To obtain a superior performance in specific tasks, deep learning methods require massive data to train its models and learn features' representation. The data for training, however, may be users' personal, highly sensitive data such as photos and voice recordings. Traditional centralized methods require users' data to be stored in companies that collect it, or users would not enjoy smart services powered by deep learning methods since they only run in the core of the network. Federated learning is dedicated to mitigating the privacy issue in the above key performance indicators. Federated learning is an emerging yet promising approach to preserve privacy when training the DNN model based on data originated by multiple clients. Rather than aggregating the raw data to a centralized datacenter for training, federated learning [McMahan et al., 2016] leaves the raw data distributed on the clients (e.g., mobile devices),

Table 2.1: Existing technologies for edge learning (*Continues.*)

Technology	Highlights	Related Work
Federated Learning	• Leave the training data distributed on the end devices • Train the shared model on the server by aggregating locally computed updates • Preserve privacy	Chen et al. (2019), Kim et al. (2018), Konečnỳ et al. (2016), Lalitha et al., (2018), McMahan et al. (2016), Shokri and Shmatikov (2015)
Aggregation Frequency Control	• Determine the best trade-off between local update and global parameter aggregation under a given resource budget • Intelligent communication control	Hsieh et al. (2017), Nishio and Yonetani (2018), Wang et al. (2019a)
Gradient Compression	• Gradient quantization by quantizing each element of gradient vectors to a finite-bit low-precision value • Gradient sparsification by transmitting only some values of the gradient vectors	Amiri and Gunduz (2019), Lin et al. (2017), Stich et al. (2018), Tang et al. (2018), Tao and Li (2018)
DNN Splitting	• Select a splitting point to reduce latency as much as possible • Preserve privacy	Harlap et al. (2018), Mao et al. (2018b), Osia et al. (2017), Wang et al. (2018b)
Knowledge Transfer Learning	• First train a base network (teacher network) on a base dataset and task and then transfer the learned features to a second target network (student network) to be trained on a target dataset and task • The transition from generality to specificity	Osia et al. (2017), Sharma et al. (2018), Wang et al. (2018b)
Gossip Training	• Random gossip communication among devices • Full asynchronization and total decentralization • Preserve privacy	Blot et al. (2016), Boyd et al. (2006), Daily et al. (2018), Jin et al. (2016)

Table 2.1: (*Continued.*) Existing technologies for edge learning

Edge-Cloud Collaboration	• Utilize the cloud data center to tackle the performance limitation of edge devices • Efficient communication process • Preserve privacy	Gao et al. (2008), Jeong et al. (2018a), Omidshafiei et al. (2017)
Meta-Learning	• Optimize for the ability to learn how to learn by exploiting correlated tasks • Enable fast learning with small sample size	Finn et al. (2017), Mishra et al. (2017), Snell et al. (2017)

and trains a shared model on the server by aggregating locally computed updates. Some main challenges of federated learning are related to optimization and communication.

For the optimization perspective, one main challenge is to optimize the gradient information for a shared model by using the distributed gradient updates on mobile devices. More specifically, federated learning adopts SGD, which updates the gradient over small subsets (mini-batch) of data samples. Shokri and Shmatikov [2015] design a selective stochastic gradient descent (SSGD) protocol, allowing the clients to train independently on their own datasets and selectively share small subsets of their models' key parameters to the centralized aggregator. Since SGD is easy to be parallelized as well as asynchronously executed, SSGD targets both privacy and training loss. Specifically, while preserving clients' own privacy, the training loss can be reduced by sharing the models among clients, comparing to training solely on their own inputs. A potential shortcoming of Shokri and Shmatikov [2015] is that it does not consider unbalanced and non-IID (none Independent Identical Distribution) data. As an extension, McMahan et al. [2016] advocate a decentralized approach, termed as federated learning, and present FedAvg method for federated learning with the DNN based on iterative model averaging. Here the iterative model averaging means that the clients update the model locally with one-step SGD and then the server averages the resulting models with weights. The optimization on McMahan et al. [2016] emphasizes the properties of unbalanced and non-IID since the distributed data may come from various sources.

Recently, personalized federated learning has emerged quickly as an promising solution to deal with the statistical heterogeneity due to the non-IID data in federated learning. Notably, a personalized variant of federated learning based on meta-learning [Finn et al., 2017] has been proposed in Lin et al. [2020] and Fallah et al. [2019], where the objective is to learn a good global model initialization such that maximal performance can be obtained with the model parameters updated with only a few data samples at a new client. By using the Moreau envelop

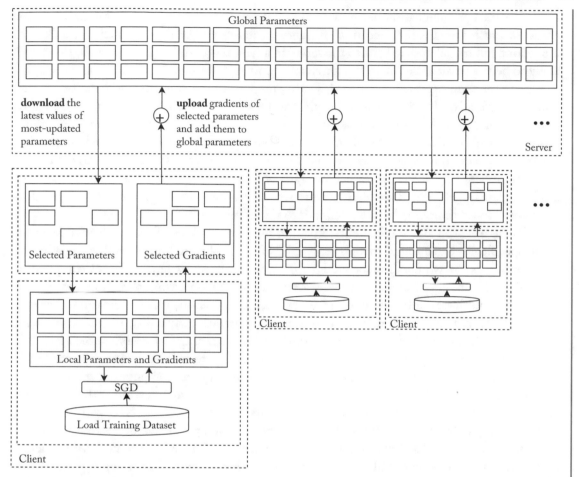

Figure 2.2: The architecture of the deep learning system based on SSGD protocol.

as a regularized loss function at each client, Dinh et al. [2020] consider a personalized federated learning algorithm by decoupling the personalized model optimization from the global model learning and formulating the learning problem as a bi-level optimization problem. Similarly, based on a rigorous characterization of the conditions when the clients can be incentivized to participate into federated learning and when the personalized model is better than the global model, an adaptive personalized federated learning approach is proposed in Deng et al. [2020]. The performance of personalized federated learning highly can depend on the effectiveness of inter-client collaboration, Huang et al. [2020] leverage an attentive message passing mechanism in federated learning to adaptively discover the underlying relationship between clients, so as to encourage clients with higher similarity to collaborate more closely.

For the communication perspective, it is the unreliable and unpredictable network that poses the challenge of communication efficiency. In federated learning, each client sends a full model or a full model update back to the server in a typical round. For large models, this step is likely to be the bottleneck due to the unreliable network connections. To decrease the number of rounds for training, McMahan et al. [2016] propose to increase the computation of local updates on clients. However, it is impractical when the clients are under severe computation resources constraint. In resolve this issue, Konečný et al. [2016] propose to reduce communication cost with two new update schemes, namely structured update and sketched update. In a structured update, the model update is first learned in a restricted space parametrized using a smaller number of variables, e.g., either low-rank or a random mask. In a sketched update, a full model update is first learned and then compressed using a combination of quantization, random rotations, and subsampling, before sending it to the server.

There has been also some attention paid to the computation-efficient federated learning. More specifically, background models (e.g., DNNs) are pushed to and trained at the edge devices serving as the workers in federated learning, which however have different computation capabilities depending on the hardware resource and the dynamic workload. Note that the FL system does not have control over the edge devices. Consequently, it is more natural that each edge device would adjust the model size to fit its computational capability, instead of updating the entire model exhaustively. A resource-aware federated learning framework named ELFISH is proposed in Xu et al. [2019], where different sets of neurons will be dynamically masked in every training cycle for different edge devices depending on their computation resources profiles, and these neurons will be recovered and updated during the aggregation at the centralized server. In Ra page et al. [2020], each device employs a neural network (NN) with a topology that fits its capability, whereas part of the NNs are shared and jointly learned across all devices.

Based on Federated Learning, there are advanced variants aiming to deal with different challenges when applying to different scenarios. To further enhance the privacy, Bonawitz et al. [2017] design a protocol for secure aggregation of high-dimensional data in federated learning setting. With a constant number of rounds, the protocol has low communication overhead as well as robustness to failures, which enable Federated Learning more secure. Zhao et al. [2018a] identify the difficulty of applying federated learning and design a systematic solution, Zoo, to deal with it. Zoo supports users to easily construct, compose and deploy different machine learning models on edge computing environment and Federated Learning is included.

Though federated learning presents a new decentralized deep learning architecture, it uses a central server for aggregating local updates. Considering the scenario of training a DNN model over a fully decentralized network, i.e., a network without a central server, Lalitha et al. [2018] propose a Bayesian-based distributed algorithm, in which each device updates its belief by aggregating information from its one-hop neighbors to train a model that best fits the observations over the entire network. Furthermore, with the emerging blockchain technique, Kim et al. [2018] propose Blockchain Federated Learning (BlockFL) with the devices' model update

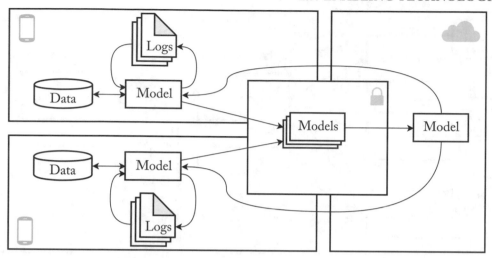

Figure 2.3: Federated learning with secure aggregation.

exchanged and verified by leveraging blockchain. BlockFL also works for a fully decentralized network, where the machine learning model can be trained without any central coordination, even when some devices lack their own training data samples.

2.3.2 AGGREGATION FREQUENCY CONTROL

This method focuses on the optimization of communication overhead during the DNN model training. On training the deep learning model in an edge computing environment, a commonly used idea (e.g., federated learning) is to train distributed models locally first, and then aggregate updates centrally. In this case, the aggregation frequency control of updates significantly impacts the communication overhead. Thus, the aggregation process, including aggregation content as well as aggregation frequency, should be designed carefully.

Based on the above insight, Hsieh et al. [2017] develop the Gaia system and the Approximate Synchronous Parallel (ASP) model for geo-distributed DNN model training. The basic idea of Gaia is to decouple the communication within a datacenter from the communication between datacenters, enabling different communication and consistency models for each (Figure 2.4). To this end, the ASP model is developed to dynamically eliminate insignificant communication between datacenters, where the aggregation frequency is controlled by the preset significance threshold (Figure 2.5). Worth noting is that Gaia focuses on geo-distributed datacenters with aboundant resources, making it potentially not amenable to edge computing nodes with constrained resources.

To incorporate the capacity constraint of edge nodes, Wang et al. [2019a] propose a control algorithm that determines the trade-off between local update and global parameter aggre-

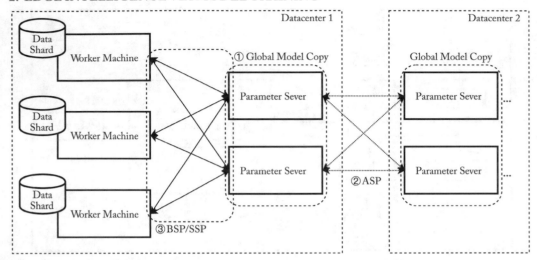

Figure 2.4: System overview of Gaia.

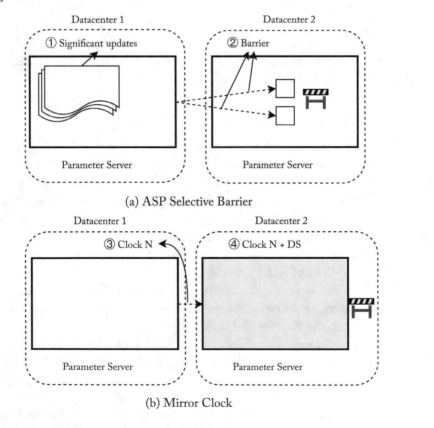

Figure 2.5: The synchronization mechanisms of ASP model.

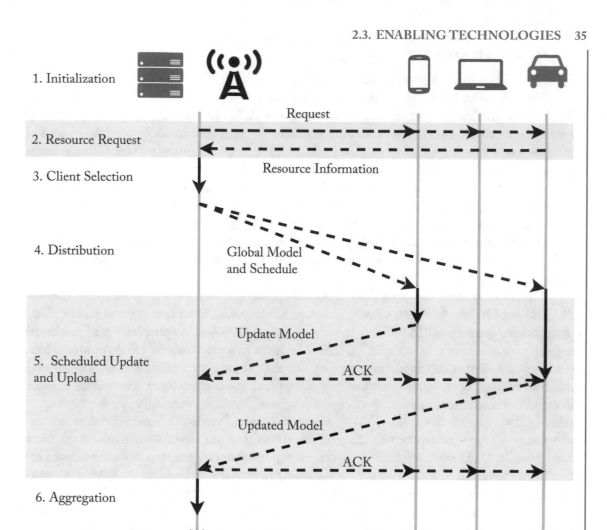

Figure 2.6: Overview of FedCS protocol.

gation under a given resource budget. The algorithm is based on the convergence analysis of distributed gradient descent and can be applied to federated learning in edge computing with provable convergence. To implement federated learning in the capacity-limited edge computing environment, Nishio and Yonetani [2018] study the client selection problem with resource constraints. In particular, an update aggregation protocol named FedCS is developed to allow the centralized server to aggregate as many client updates as possible and to accelerate performance improvement in machine learning models. An illustration of FedCS is shown in Figure 2.6.

2.3.3 GRADIENT COMPRESSION

To reduce the communication overhead incurred by decentralized training, gradient compression is another intuitive approach to compress the model update (i.e., gradient information). To this end, gradient quantization and gradient sparsification have been advocated. Specifically, gradient quantization performs lossy compression of the gradient vectors by quantizing each of their elements to a finite-bit low precision value. Gradient sparsification reduces the communication overhead by transmitting part of the gradient vectors.

Lin et al. [2017] observe that 99.9% of the gradient exchange in distributed SGD are redundant, which pinpoints to the utility of gradient compression. Based on this observation, Lin et al. propose Deep Gradient Compression (DGC), which compresses the gradient by 270–600 times for a wide range of CNNs and RNNs. To preserve accuracy during this compression, DGC employs four methods: momentum correction, local gradient clipping, momentum factor masking, and warm-up training.

Inspired by the above work [Lin et al., 2017], Tao and Li [2018] propose Edge Stochastic Gradient Descent (eSGD), a family of sparse schemes with both convergence and practical performance guarantees. To improve the first-order gradient-based optimization of stochastic objective functions in edge computing, eSGD includes two mechanisms: (1) determine which gradient coordinates are important and only transmit these coordinates; and (2) design momentum residual accumulation for tracking out-of-date residual gradient coordinates in order to avoid low convergence rate caused by sparse updates. A concise convergence analysis of sparsified SGD is given in Stich et al. [2018], where SGD is analyzed with k-sparsification or compression (e.g., top-k or random-k). The analysis shows that this scheme converges at the same rate as vanilla SGD when equipped with error compensation (keeping track of accumulated errors in memory). In other words, communication can be reduced by a factor of the dimension of the problem (sometimes even more) while still converging at the same rate.

Gradient quantization can also reduce the communication bandwidth. In a network environment, there are two common techniques to reduce communication overhead: communication compression and decentralization. Combining both techniques, In this regard, Tang et al. [2018] develop a framework of compressed, decentralized training and propose two different algorithms, called extrapolation compression and difference compression, respectively. The analysis on the two algorithms proves that both converge at the rate of $O(1/\sqrt{nT})$ where n is the number of clients and T is the number of iterations, matching the convergence rate for full precision, centralized training. Amiri and Gunduz [2019] implement distributed stochastic gradient descent (DSGD) at the wireless edge with the help of a remote parameter server. Besides, Amiri et al. further develop DSGD in digital and analog scheme, respectively. Digital DSGD (D-DSGD) assumes that the clients operate on the boundary of the Multiple Access Channel (MAC) capacity region at each iteration of DSGD algorithm, and employs gradient quantization and error accumulation to transmit their gradient estimates within the bit budget allowed by the employed power allocation. In Analog DSGD (A-DSGD), the clients first sparsify their gradient estimates with error accumulation and then project them to a lower dimensional space imposed by the available channel bandwidth. These projections are transmitted directly over the MAC without employing any digital code.

Gradients compression is a technique that involves both software and hardware. In the literature most of the attention is paid on algorithmic advancement, but Li et al. [2018e] set out to reduce communication overhead by combining both two levels. Li et al. focus on Network Interface Cards (NICs) since they are the essential carrier of communication in edge computing. To maximize the benefits of in-network acceleration, Li et al. proposed INCEPTIONN (In-Network Computing to Exchange and Process Training Information Of Neural Networks), which includes a lightweight and hardware-friendly lossy-compression algorithm for floating-point gradients, and an aggregator-free training algorithm that exchanges gradients in both legs of communication in the group. INCEPTIONN reduces the communication time by 70.9–80.7% and offers 2.2–3.1× speedup over the conventional model training while achieving the same level of accuracy.

2.3.4   DNN SPLITTING

The aim of DNN splitting is to protect privacy. DNN splitting protects user privacy by transmitting partially processed data rather than transmitting raw data. To enable a privacy-preserving edge-based training of DNN models, DNN splitting is conducted between the end devices and the edge server. This bases on the important observation that a DNN model can be split inside between two successive layers with two partitions deployed on different locations without losing accuracy.

An inevitable problem on DNN splitting is how to select the splitting point such that distributed DNN training is still under the latency requirement. On this problem, Mao et al. [2018b] utilize the differentially private mechanism and partition DNN after the first convolutional layer to minimize the cost of mobile devices. The proof in Mao et al. [2018b] guarantees that applying differentially private mechanism on activations is feasible for outsourcing training tasks to untrusted edge servers. Wang et al. [2018b] consider this problem across mobile devices and cloud datacenters. To benefit from the computation power of cloud datacenters without privacy risks, Wang et al. designed Arden (privAte infeRence framework based on DNNs), a framework which partitions the DNN model with a lightweight privacy-preserving mechanism. By data nullification and random noise addition, Arden achieves privacy protection. Considering the negative impact of private perturbation to the original data, Wang et al. use a noisy training method to enhance the cloud-side network robustness to perturbed data.

Osia et al. [2017] introduce a hybrid user-cloud framework on the privacy issue, which utilizes a private-feature extractor as its core component and breaks down large, complex deep models for cooperative, privacy-preserving analytics. In this framework, the feature extractor module is properly designed to output the private feature constrained to keeping the primary information while discarding all the other sensitive information. Three different techniques are

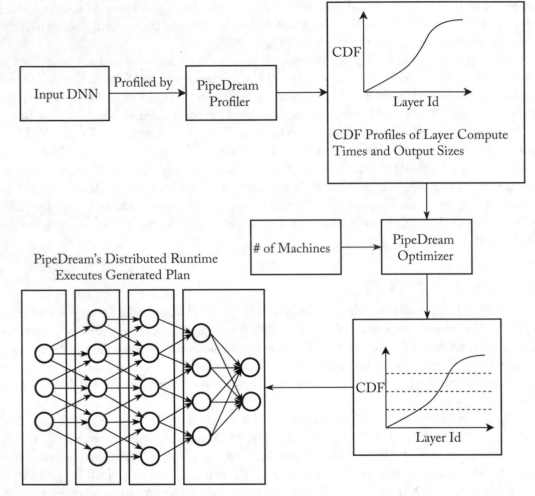

Figure 2.7: PipeDream's automated mechanism.

employed to ensure sensitive measures: dimensionality reduction, noise addition, and Siamese fine-tuning.

When applying DNN splitting to privacy-preserving, it is remarkable that this technique also works for dealing with the tremendous computation of DNN. Exploiting the fact that edge computing usually involves a large number of devices, parallelization approaches is usually employed to manage DNN computation. DNN training in parallel includes two kinds of parallelism, data parallelism and model parallelism. However, data parallelism may bring heavy overhead of communication while model parallelism usually leads to severe under-utilization of computation resources. To address these problems, Harlap et al. [2018] propose pipeline parallelism, an enhancement to model-parallelism, where multiple mini-batches are injected into the system at once to ensure efficient and concurrent use of computation resources. Based on pipeline parallelism, Harlap et al. design PipeDream, a system which supports pipelined training, and automatically determines how to systematically split a given model across the available computing nodes. PipeDream shows the advantage of reducing communication overhead and utilizing computing resource efficiently. The overview of PipeDream's automated mechanism is in Figure 2.7.

2.3.5   KNOWLEDGE TRANSFER LEARNING

Knowledge transfer learning, or transfer learning for simplicity, is closely connected with DNN splitting technique. In transfer learning, for the purpose of reducing DNN model training energy cost on edge devices, we first train a base network (teacher/mentor network) on a base dataset, and then we repurpose the learned features, i.e., transfer them to a second target network (student/mentee network) to be trained on a target dataset. This process can work well if the features are general (i.e., suitable to both base and target tasks) instead of specific to the base task. The transition involves a process from generality to specificity.

Yosinski et al. [2014] give an experimental quantification on the generality vs. specificity of neurons in each layer of a CNN. By a series of experimental evaluations, Yosinski et al. draw that transferability is negatively affected by two distinct issues: (1) the specialization of higher layer neurons to their original task at the expense of the performance loss of the target task, which was expected; and (2) optimization difficulties related to splitting networks between co-adapted neurons, which is not expected. Another interesting result is that initializing a network with transferred features from almost any number of layers can produce a boost to the generalization that lingers even after fine-tuning of the target dataset. This showcases the potential utility of transfer learning, thereby spurring the research on the synergy between transfer learning and edge computing.

Transfer learning is appealing for learning on edge devices since it can greatly reduce resource demand, but a thorough investigation on its effectiveness is lacking. To bridge this gap, Sharma et al. [2018] and Chen et al. [2019] provide extensive studies on the performance (in both accuracy and convergence speed) of transfer learning, considering different student network architectures and different techniques for transferring knowledge from teacher to student. The result varies with architectures and transfer techniques. Different types of knowledge transfer techniques are shown in Figure 2.8. Noticeable performance improvement is obtained by transferring knowledge from both the intermediate layers and last layer of the teacher to a shallower student neural network while other architectures and transfer techniques do not fare so well and some of them even lead to negative performance impact.

Transfer learning treats the shallow layers of a pre-trained DNN on one dataset as a generic feature extractor that can be applied to other target tasks or datasets. With this feature, transfer learning is employed in many pieces of research and inspires the design of some frameworks.

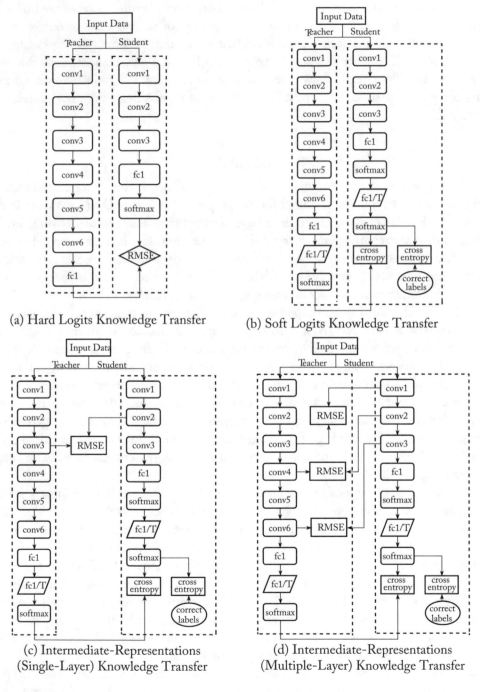

(a) Hard Logits Knowledge Transfer

(b) Soft Logits Knowledge Transfer

(c) Intermediate-Representations
(Single-Layer) Knowledge Transfer

(d) Intermediate-Representations
(Multiple-Layer) Knowledge Transfer

Figure 2.8: Different kinds of knowledge transferring techniques.

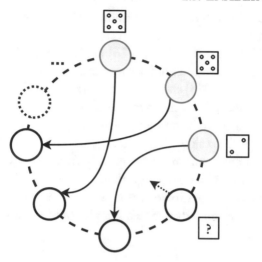

Figure 2.9: Communication with randomly selected partner in gossip manner.

Osia et al. [2017], which we have alluded to in Section 2.3.4, use transfer learning to determine the degree of generality and particularity of a private feature. Arden, proposed in Wang et al. [2018b], partitions a DNN across the mobile device and the cloud data center, where the raw data is transformed by the shallow portions of the DNN on the mobile device side. As Wang et al. [2018b] referred, the design of DNN splitting in Arden is inspired by transfer learning.

2.3.6   GOSSIP TRAINING

Aiming at shortening the training latency, gossip training is a new decentralized training method built on randomized gossip algorithms. Early work on random gossip algorithms [Boyd et al., 2006] uses gossip averaging which can converge fast toward a consensus by information exchange among peers. The gossip distributed algorithms enjoy the advantage of full asynchronization and total decentralization as they have no requirement on centralized nodes or variables. Inspired by this, Gossip Stochastic Gradient Descent (GoSGD) Blot et al. [2016] proposed to train DNN models in an asynchronous and decentralized way. GoSGD manages a group of independent nodes, where each of them hosts a DNN model and iteratively proceeds two steps: gradient update and mixing update. Specifically, each node updates its hosted DNN model locally in the gradient update step and then shares its information with another randomly selected node in mixing update step, as shown in Figure 2.9. The steps repeat until all the DNNs converge on a consensus.

The aim of GoSGD is to speed up the training of convolutional networks. Another gossip-based algorithm, namely gossiping SGD [Jin et al., 2016], is designed to retain the positive features of both synchronous and asynchronous SGD methods. Gossiping SGD replaces the all-reduced collective operation of synchronous training with a gossip aggregation algorithm, in an asynchronous manner.

Both Blot et al. [2016] and Jin et al. [2016] apply gossip algorithms on the updates of SGD, but neither of them are implemented at large scale. In contrast, with the deployment in large-scale systems, Daily et al. [2018] show that the naive gossip-based algorithms at scale lead to a communication imbalance, poor convergence and heavy communication overhead. To mitigate these issues, Daily et al. introduce GossipGraD, a gossip communication efficient SGD algorithm which is practical for scaling deep learning algorithms on large scale systems. GossipGrad reduces the overall communication complexity from $\Theta(\log(p))$ to $O(1)$ and considers diffusion such that computing nodes exchange their updates (gradients) indirectly after every $\log(p)$ steps. It also considers the rotation of communication partners for facilitating direct diffusion of gradients and asynchronous distributed sample shuffling during the feedforward phase in SGD to prevent over-fitting.

2.3.7   HARDWARE ACCELERATION

Since simply using software methods such as gradient compression techniques may fail to satisfy some stringent latency requirements at the edge, hardware acceleration for distributed DNN training would be useful. In DNN models, convolutional layers and fully connected layers consume a majority of computation [Sze et al., 2017]. The two kinds of layers share the same arithmetical computation, the MAC operations, which can be easily parallelized. Based on this important fact, highly-parallel computing paradigms are adapted on hardware design to accelerate DNN training.

We elaborate on two kinds of hardware architectures as shown in Figure 2.10. The first one is the temporal architecture in Figure 2.10a, which employs a variety of techniques to improve parallelisms such as vectors (SIMD) and parallel threads (SIMT). The temporal architecture appears mostly in CPUs and GPUs, which has been the fundamental platforms for DNN training. The second one is the spatial architecture in Figure 2.10b, commonly used in ASIC- and FPGA-based designs. Thanks to the flexibility and universality of ASIC and FPGA, mostly newly designed hardware is based on this spatial architecture currently.

An instance based on ASIC and FPGA is ScaleDeep [Venkataramani et al., 2017], proposed by Venkataramani et al. ScaleDeep is a dense, scalable, and spatial architecture primarily targeting DNN training, which is composed by subsystems that are specialized to leverage the computation and communication characteristics of DNNs. ScaleDeep derives its efficiency with key architectural features as follows: (1) heterogeneous processing tiles and chips, (2) a memory hierarchy and 3-tiered interconnect topology, (3) a low-overhead synchronization mechanism, and (4) methods to map DNNs to the proposed architecture. Furthermore, Venkataramani et al. develop a compiler to allow programming any DNN topology onto ScaleDeep, and a detailed architectural simulator to estimate performance and energy.

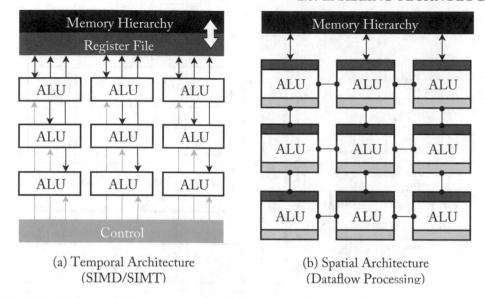

(a) Temporal Architecture
(SIMD/SIMT)

(b) Spatial Architecture
(Dataflow Processing)

Figure 2.10: Highly parallel computing paradigms.

Wang et al. [2016] propose another architecture based on FPGA, which accelerates DNN training by approximate arithmetic instead. They further present an implementation of the Liquid State Machine, a spiking neural network model for real-world pattern recognition problems. The proposed architecture of Wang et al. [2016] consists of a parallel digital reservoir with fixed synapses, and a readout stage that is tuned by a biologically plausible supervised learning rule. On the same issue, Wang et al. [2017a] present the parallel neuromorphic processor architectures for spiking neural networks, as shown in Figure 2.11. The proposed architectures of Wang et al. [2017a] address several critical issues pertaining to efficient parallelization of the update of membrane potentials, on-chip storage of synaptic weights, and integration of approximate arithmetic units.

Hardware design with ASIC or FPGA is challenging, which requires long design cycles and extensive expertise. To tackle this challenge, Mahajan et al. [2016] develop Tabla, a framework that generates the synthesizable implementation of the accelerator for FPGA realization for a class of machine learning algorithms. Regarding machine learning algorithms as stochastic optimization problems, Tabla identifies the commonalities across them and utilizes this commonality to provide a high-level abstraction for programmers. The overview of the workflow with Tabla is shown in Figure 2.12. Similarly, Park et al. [2017] offer a full computing stack constituting language, compiler, system software, template, named CoSMIC. CoSMIC enables programmers to exploit scale-out acceleration using multiple FPGAs and Programmable ASICs (PASICs) from a high-level and mathematical Domain-Specific Language (DSL).

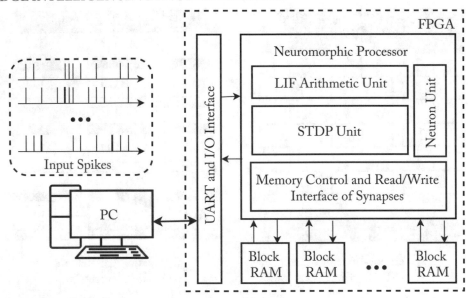

Figure 2.11: Top-level schematic of the proposed neuromorphic processor.

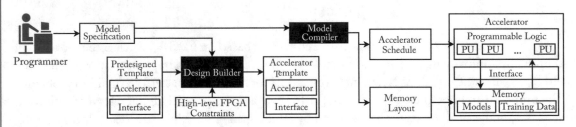

Figure 2.12: The overview of the workflow with Tabla.

It is remarkable that the Cloud TPU [J. Dean, U. Hölzle, 2019], released by Google, is also based on ASIC. Google Cloud TPU is the next-generation TPU which is purpose-built to accelerate the Google's distributed machine learning framework, TensorFlow [Abadi et al., 2016]. For DNN computation at the edge, Edge TPU is introduced with Cloud IoT Edge, where Edge TPU is a new hardware chip specialized for edge computing and Cloud IoT Edge is a software stack that extends the cloud capability to the edge. A further investigation of the marketplace of Google TPU is given in Setion 8.2.

2.3.8 KNOLWEDGE DISTILLATION FOR EDGE-CLOUD COLLABORATION

Training models directly on edge devices based on the local dataset, if not designed intelligently, would suffer from poor performance and energy inefficiency. One of main reasons is that the AI model training process requires high computational power and is even more energy-intensive for some real-time applications, which indicates that most of edge devices may not be capable to learn sophisticated AI models solely with their limited computing power. Another key reason is that the amount of local data samples available at each edge IoT device is often limited. Based on these limited local datasets only, it would be difficult for edge nodes to obtain an accurate predictive model, especially for sophisticated deep learning and RL algorithms, since model training phases require large amounts of data samples. It is therefore of great importance to build an innovative edge-cloud collaboration to boost up the edge learning performance.

During collaborative learning between cloud and edge, it is assumed that the cloud data center has access to large amount of historical data and plenteous computing resources. Thus, the cloud has the ability to learn an AI model with enough accuracy. To enable the innovative collaboration between edge and cloud, one of technologies for EI models training is parameter-based transfer learning. Specifically, the transferred knowledge is encoded into parameters with some specific parametric models and a large number of samples are required in the cloud data center to accurately estimate the parameters that need to be transferred. A popular approach is the Bayesian learning framework, where some prior knowledge is encoded into a prior distribution with specific model parameters. For example, Chelba and Acero [2006] considered this method for the task of adapting a maximum entropy capitalizer across domains. Li and Bilmes [2007] also proposed a general Bayesian divergence prior framework for domain adaptation. We note that a further investigation of hierarchical Bayesian learning framework will be given in Section 4.3 for EI applications. Besides the construction of Bayesian prior, Gao et al. [2008] proposed a dynamically weighted ensemble learning framework, which can combine multiple models according to their different prediction power.

Needless to say, downloading pre-trained AI model parameters from the cloud can help edge devices to tackle challenges induced by limited power and scarce training samples. However, direct transmission of model parameters may entail communication overhead being proportional to model sizes. To mitigate this communication inefficiency, knowledge distillation [Hinton et al., 2015] (KD) approach is proposed and widely used in edge-cloud collaboration model training process, especially for DNN training. KD focuses on training a student NN, which has a compressed model size and simpler architecture, based on a well-trained teacher NN. The key idea of KD is adding a distillation regularizer to the student NN's loss function, where this extra regularizer is formulated by cross entropy to measure the knowledge gap between the teacher and the student. Instead of exchanging the model parameters (i.e., weights or gradients), KD transmits the model output information between the cloud and edge devices. More specifically, the model output information here is normalized logits whose payload size depends only on the number of labels, which is far less than the dimension of highly complex DNN model parameters. Omidshafiei et al. [2017] presented a KD-based approach for distilling single-task policies into a unified policy that performs well across multiple related tasks. Following the same spirit of federated learning, Jeong et al. [2018a] proposed federated distillation, where normalized logits are uploaded from each edge device to the cloud server and edge devices periodically download averaged global normalized logits for local training. Besides matching logits between the cloud and the edge, model input-output Jacobian matching is proposed in Srinivas and Fleuret [2018] and is intepreted as a KD process that inserts noise into inputs.

2.3.9   META-LEARNING

Meta-learning [Thrun and Pratt, 2012] has recently emerged again as a promising solution for fast learning with small sample size, by developing new tools that can take advantage of the power of the latest neural architectures. Different with traditional transfer learning, meta-learning explicitly optimizes for the ability to learn how to learn by exploiting the prior experience from related tasks, which enables fast learning over unseen tasks using only a few samples from those tasks.

In general, current meta-learning algorithms can be broadly grouped into three different categories: (1) *Gradient-based methods:* use gradient descent to adapt the model parameters using training samples [Finn et al., 2017, Ravi and Larochelle, 2017]; (2) *Model-based methods:* learn a parameterized predictor, e.g., a recurrent network, to estimate the model parameters [Mishra et al., 2017, Munkhdalai et al., 2017]; and (3) *Metric-learning based methods:* learn a distance-based prediction rule over the embeddings [Snell et al., 2017, Vinyals et al., 2016]. Among these, a gradient-based method named MAML [Finn et al., 2017] is particularly simple and effective which assumes that all tasks can be represented by the same class of parameterized model. Specifically, MAML directly optimizes the learning performance with respect to an initialization of the model such that even one-step gradient descent from that initialization can still produce good results on a new task with only a few data samples from that task. A number of recent papers have studied and extended this approach [Antoniou et al., 2018, Grant et al., 2018, Li et al., 2017]. In particular, Nichol et al. [2018] propose a first-order method named Reptile, which is similar to joint training but surprisingly works well a meta-learning algorithm to circumvent the need of the second derivatives in MAML. These approaches have been extended to devise new reinforcement learning algorithms, which perform significantly better than standard reinforcement learning algorithms that learn from scratch [Gupta et al., 2018, Nagabandi et al., 2018, Rakelly et al., 2019].

The advantages of meta-learning make it naturally a promising technique to achieve real-time edge intelligence from the following several aspects: (1) *Fast learning:* many IoT applications at the network edge, such as autonomous driving and augmented reality, need intensive computation to accomplish object tracking, content analytics and intelligent decision in a real-time manner, in order to meet the requirements for safety, accuracy, performance and user experience. (2) *Small sample size:* while tremendous data will be generated by the ensemble of IoT devices, the amount of personal data at every edge node is limited. So the techniques applied at the edge have to be able to achieve good learning performance with a small amount of local data. (3) *Task generation:* by making use of the fact that different edge nodes often have distinct local models while sharing some similarity, and are continuously making intelligent decisions, the tasks can be automatically constructed across different edge nodes and are naturally correlated. This also motivates a platform-aided collaborative learning framework where the model knowledge is learned collaboratively by a federation of edge nodes in a distributed manner, and then is transferred via the platform to the target node for fine-tuning with its local dataset, to enable real-time edge intelligence in a distributed manner.

2.4   SUMMARY

In this chapter, we present an overview of the systems and frameworks for distributed EI model training on the edge, as outlined in Table 2.2, including the architecture, EI level, objectives, employed technologies, and effectiveness.

High-performance, low-resource consumption, and robustness are the common pursuit of those systems and frameworks as listed in Table 2.2. Specifically, high performance represents the low loss function value that the DNN model achieves by distributed training. To evaluate the performance more accurately, the influence factors, including the whole latency, the testing DNN model and the available computation resource, are fixed in the experimental setup. Low-resource consumption is a board statement, where the resources range from network bandwidth, energy power to computation capability. We expect the system or the framework to consume as little as resources, since it significantly influences the total economic viability of distributed DNN training, especially for commercial companies. Robustness implies the ability that the system or the framework cope with errors. A robust system is able to identify invalid input and properly deal with it. In distributed DNN training, robustness requires the system to train data with special properties such as non-IID, perturbance, and noise. Robustness is practically significant for commercial products since users' inputs are often versatile and noisy.

In general, a key challenge for distributed EI model training is the data privacy issue. It is because the distributed data sources may originate from individual persons and different organizations. For users, they may be sensitive to their own private data, not allowing any private information to be shared. For companies, they have to consider the privacy policy to avoid legal subpoenas and extra-judicial surveillance. Therefore, the design of distributed training systems needs to carefully consider privacy preservation. Systems considering privacy issues in Table 2.2 include FedAvg, BlockFL, GossipGraD, and so on. The decentralized architecture is naturally friendly to users' privacy, for which the systems that are based on the decentralized architecture such as BlockFL and GossipGraD typically preserve privacy better. As a contrast, the centralized architecture involves a centralized data collection operation, and the hybrid architecture requires a data transmission operation. For this reason, the systems based on these two architectures would implement more extra efforts in data privacy protection.

Compared with the DNN training under the cloud-based framework, the DNN training in an edge-based framework pays more attention to protecting users' privacy and training an available deep learning model faster. Under cloud-based training, a large amount of raw data

Table 2.2: An overview of systems and frameworks on EI model training (*Continues.*)

System or Framework	Architecture	EI Level	Objectives	Employed Technology	Effectiveness
FedAvg (McMahan et al., 2016)	Hybrid	Level-4	• Robustness to non-IID and unbalanced optimization • Low communication cost • Privacy preservation	• Federated Learning • Iterative model averaging	• Reduce communication rounds by 10–100× as compared to synchronized stochastic gradient descent
SSGD (Shokri and Shmatikov, 2015)	Hybrid	Level-4	• Jointly training a DNN model among clients • Privacy preservation	• Federated Learning • Selective SGD	• Clients' privacy is preserved while the model accuracy beyond training solely
Zoo (Zhao et al. 2018a)	Hybrid	Level-4	• Reducing communication cost • Privacy preservation	• Federated Learning • Composable services	• Processes each image within constant time despite the size difference of images
BlockFL (Kim et al., 2018)	Decentralized	Level-6	• Federated Learning in decentralized manner • Low latency • Privacy preservation	• Federated Learning • Blockchain	• Latency increase up to 1.5% to achieve the optimal block generation than the simulated minimum latency
Gaia (Hsieh et al., 2017)	Centralized	Cloud Intelligence	• Geo-distributed scalability • Intelligent communication mechanism over WANs • Generic and flexible for most machine learning algorithms	• Aggregation frequency control • ASP model	• Speedup 1.8–53.5× over distributed machine-learning systems • Within 0.94–1.40× of the speed of running the same machine-learning algorithm on machines on a local area network (LAN)
DGC (Lin et al., 2017)	N/A	N/A	• Reducing the communication bandwidth • High compression rate without losing model accuracy • Fast convergence	• Gradient Compression • Momentum correction • Local gradient clipping • Momentum factor Masking • Warm-up training	• Achieve a gradient compression ratio from 270–600× without losing accuracy • Cut the gradient size of ResNet-50 from 97–0.35 MB and for DeepSpeech from 488–0.74 MB

Table 2.2: (*Continued.*) An overview of systems and frameworks on EI model training (*Continues.*)

eSGD (Tao and Li 2018)	Hybrid	Level-4	• Scaling up edge training of CNN • Reducing communication cost	• Selective transmit important gradient coordinates • Momentum residual accumulation	• Reach 91.2%, 86.7%, 81.5% accuracy on MNIST data set with gradient drop ratio 50%, 75%, 87.5%, respectively
INCEPTIONN (Li et al., 2018e)	Hybrid	Level-5	• Maximizing the opportunities for compression • Avoiding the bottleneck at aggregators	• Lossy gradient compression NICintegrated compression accelerator • Gradient-centric aggregator-free training	• Reduce the communication time by 70.9–80.7% • Offer 2.2–3.1× speedup over the conventional training system while achieving the same level of accuracy
Arden (Wang et al., 2018b)	Centralized	Cloud Intelligence	• Maximize utilization of computing resources • Low latency • Fast convergence	• DNN splitting • Arbitrary data nullification • Random noise addition	• The average reductions compared with the other four DNNs in terms of time, memory, and energy are 60.10%, 92.07%, and 77.05%, respectively
PipeDream (Harlap et al., 2018)	Hybrid	Level-5	• Maximizing utilization of computing resources • Low latency • Fast convergence	• DNN splitting • Pipeline parallelis	• Using 4 machines to train the > 100 million parameter VGG16 on the ImageNet1K dataset, PipeDream converges 2.5× faster than using a single machine and 3× faster than data parallel training
GoSGD (Blot et al., 2016)	Decentralized	Level-6	• Speeding up DNN training • Fast convergence	• Gossip training	• Make better use of the exchanges comparing to EASGD • Converge a lot faster compar-

Table 2.2: (*Continued.*) An overview of systems and frameworks on EI model training

Gossiping SGD (Jin et al., 2016)	Decentralized	Level-6	• Speeding up DNN training • Scaling up DNN training • Asynchronous training	• Gossip training • Model partition	• One iteration of gossiping SGD is faster than one iteration of all-reduce SGD • Work quickly at the initial step size
GossipGraD (Daily et al., 2018)	Decentralized	Level-6	• Reducing communication complexity • Fast convergence • Privacy preservation	• Gossip training • Model partition	• Achieve about 100% compute efficiency for ResNet50 using 128 NVIDIA Pascal P100 GPUs while matching the top-1 classification accuracy published in literature

generated at the client side is directly transmitted to the cloud data center through the long WAN, which not only causes hidden dangers of user privacy leakage but also consumes huge bandwidth resources. Moreover, in some scenarios such as military and disaster applications when the access to the cloud center is impossible, the edge-based training will be highly desirable. On the other hand, the cloud data center can collect a larger amount of data and train an AI model with more powerful resources, and hence the advantage of cloud intelligence is that it can train a much larger-scale and more accurate model.

Convergence is another important metric to be considered. This metric is specialized for systems using decentralized methods and parallel methods. The decentralized method, such as gossip-based methods, train DNN models by iteratively updating the gradients on devices without aggregating them. The parallel methods parallelize training data or DNN model on multiple threads. These methods make sense if the updates or the threads converge to a consensus. For example, there are two systems, PipeDream and GossipGraD, set out to achieve fast convergence in Table 2.2. PipeDream is the system based on the hybrid architecture which employs pipeline parallelism and GossipGraD scales up distributed DNN training based on decentralized gossip training.

CHAPTER 3

Edge Intelligence via Federated Meta-Learning

As noted before, it is anticipated that a high percentage of IoT data will be stored and processed locally. However, because many AI applications typically require high computational power that greatly outweighs the capacity of resource- and energy-constrained IoT devices, it is highly challenging for a single edge node alone to achieve real-time edge intelligence, which points to the need of collaborative learning that is capable of leveraging the knowledge transferred from other edge nodes or the cloud. In this chapter, we focus on collaborative learning across edge nodes, and turn our attention to collaborative learning between the edge and the cloud in next chapter, aiming to fully leverage the potentially valuable knowledge transfer from the cloud.

3.1   INTRODUCTION

Based on the key observation that learning tasks across edge nodes often share some similarity, we propose a platform-aided collaborative learning framework where the model knowledge is learned collaboratively by a federation of edge nodes, in a distributed manner, and then is transferred via the platform to the target edge node for fine-tuning with its local dataset. Then, a key question to ask is *"What knowledge should the federation of edge nodes learn and be transferred to the target edge node for achieving real-time edge intelligence?"*

Federated learning has recently been developed for model training across multiple edge nodes, where a single global model is trained across all edge nodes in a distributed manner. It has been shown that limited performance is achieved when fine-tuning the global model for adaptation to a new (target) edge node with a small dataset [Ravi and Larochelle, 2017]. Along a different line, federated multi-task learning [Smith et al., 2017] has been proposed to train different but related models for different nodes, aiming to deal with the model heterogeneity among edge nodes. In particular, every source edge node, which is also a target edge node, is able to learn a unique model through capitalizing the computational resource and data belonging to other nodes. This, however, inevitably requires intensive computation and communications and is time-consuming, and hence could not meet the latency requirement for real-time edge intelligence.

Building on the recent exciting advances in meta-learning [Finn et al., 2017, Schmidhuber, 1987], in this chapter, we first propose a platform-aided collaborative learning framework where a model is first trained by a federated meta-learning (FedML) approach across multiple edge nodes, and then is transferred via the platform to the target edge node such that rapid adaptation can be achieved with the model updated (e.g., through one gradient step) with small local datasets, in order to achieve real-time edge intelligence. Next, we study the convergence behavior of the FedML algorithm and examine the adaptation performance of the fine-tuned model at the target edge node. To establish the convergence, we impose bounds on the variations of the gradients and Hessians of local loss functions (with respect to the hyper-parameter) across edge nodes, thereby removing the assumption in meta-learning that requires all tasks follow a (known) distribution. To combat against the possible vulnerability of meta-learning algorithms, we propose a robust version of FedML, building on recent advances in distributionally robust optimization (DRO). We further show that the proposed algorithm still converges under mild technical conditions. In the end, we evaluate the performance of the proposed collaborative learning framework using different datasets, which corroborates the effectiveness of FedML and showcases the robustness of the DRO-based robust FedML.

Different from meta-learning which requires the knowledge of the task distribution [Finn et al., 2017], the FedML proposed in this work removes this assumption, by making use of the fact that different edge nodes often have distinct local models while sharing some similarity, and it can automate the process of task construction because each edge node is continuously making intelligent decisions. Further, the FedML eliminates the need for centralized computation and thus offers the flexibility to strike a good balance between the communication cost and local computation cost (e.g., via controlling the number of local update steps).

3.2   RELATED WORK

To the best of our knowledge, our work Lin et al. [2020] is among the first to establish the convergence of (federated) meta-learning. During the preparation of this monograph, the preprint of one concurrent work [Fallah et al., 2019] about convergence analysis of meta-learning algorithms became available online. It is worth noting that Fallah et al. [2019] studies the convergence of centralized MAML algorithms for non-convex functions, whereas this chapter focuses on the convergence and adaptation performance of the FedML algorithm with node similarity to achieve real-time edge intelligence in a federated setting, where multiple local update steps are allowed to balance the trade-off between the communication cost and local computation cost. A similar meta-learning framework is studied in Chen et al. [2018a] for recommendation systems through assigning one task to every user, which nevertheless does not consider any system modeling in federated learning.

The susceptibility of meta-learning algorithms such as MAML to adversarial attacks is first investigated in Edmunds et al. [2017], and recently Yin et al. [2018] also demonstrates the significant performance degradation of MAML with adversarial samples. To make meta-learning more robust, Yin et al. [2018] proposes a meta-learning algorithm called ADML which exploits both clean and adversarial samples to push the inner gradient update to arm-wrestle with the meta-update. Unfortunately, this type of approaches are generally intractable. The DRO-based robust FedML algorithm proposed here is not only computationally tractable, but also resistant to more general perturbations, e.g., out-of-distribution samples. In addition, the trade-off between robustness and accuracy can be fine-tuned by the size of the distributional uncertainty set.

Federated learning has recently attracted much attention (see e.g., McMahan et al. [2016], Sahu et al. [2018], Wang et al. [2019a]), which leaves the training data distributed on the nodes, and trains a shared model on the server by aggregating locally computed updates. Notwithstanding, federated learning is not designed for fast learning with small datasets. In particular, federated learning intends to find a global model that fits the data as accurately as possible for all participating nodes, wheras federated meta-learning learns a model initialization, from which fast adaptation from even small datasets can still reach good performance, and also keeps the node heterogeneity in the sense that different models would be learned for different nodes after quick adaptation from the global model initialization.

Both meta-learning and multi-task learning [Evgeniou and Pontil, 2004, Ruder, 2017, Smith et al., 2017] aim to improve the learning performance by leveraging other related tasks. However, meta-learning focuses on the fast learning ability with small sample sizes and the performance improvement at the target (learning at the source is irrelevant), whereas multi-task learning aims to learn both the source and target tasks simultaneously and accurately. Besides, the model initialization learned by meta-learning can be fine-tuned with good performance on various target tasks using minimal data points, while multi-task learning may favor tasks with significantly larger amount of samples than others.

3.3   PRELIMINARIES ON META-LEARNING

In standard meta-learning, all the tasks are assumed to follow the same distribution, $\tilde{\mathcal{T}} \sim \mathbb{P}(\tilde{\mathcal{T}})$. At the meta-training time, a set of J tasks $\{\tilde{\mathcal{T}}_j\}_{j=1}^{J}$ are drawn from the prior distribution $\mathbb{P}(\tilde{\mathcal{T}})$ with the corresponding dataset $\mathcal{D}_j$. At the meta-testing time, we are facing a new task $\tilde{\mathcal{T}}_i \sim \mathbb{P}(\tilde{\mathcal{T}})$ with a small dataset $\mathcal{D}_i$. Meta-learning aims to find a model based on the J tasks sampled at the training time such that the model can be optimal for the testing task $\tilde{\mathcal{T}}_i$ after rapid adaptation.

In MAML, the tasks are assumed to follow a meta-model which is represented by a parameterized function f_θ with parameter θ. For each training task $\tilde{\mathcal{T}}_j$, its dataset $\mathcal{D}_j$ is divided into two separate sets, the training set $\mathcal{D}_j^{train}$ and the testing set $\mathcal{D}_j^{test}$. When adapting to task $\tilde{\mathcal{T}}_j$, the model's parameters θ become ϕ_j. The updated parameter ϕ_j can be computed using one or more gradient descent steps with the training set $\mathcal{D}_j^{train}$, e.g., for a given loss function $\mathcal{L}$ and learning rate α,

$$\phi_j = \theta - \alpha \nabla_\theta \mathcal{L}(\mathcal{D}_j^{train}, \theta).$$

MAML directly optimizes for the parameter θ such that the updated parameter ϕ_i with small dataset $\mathcal{D}_i$ is optimal for a new task $\tilde{\mathcal{T}}_i$. Specifically, MAML solves the following optimization

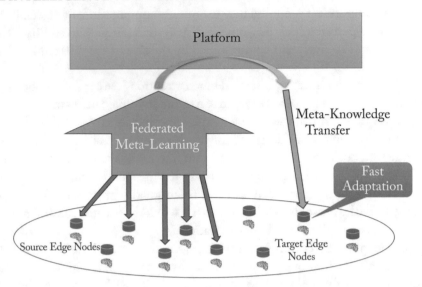

Figure 3.1: A platform-aided collaborative learning framework with FedML for real-time edge intelligence.

problem:

$$\min_{\theta} \sum_{\tilde{\mathcal{T}}_j \sim \mathbb{P}(\tilde{\mathcal{T}})} \mathcal{L}(\mathcal{D}_j^{test}, \theta - \alpha \nabla_\theta \mathcal{L}(\mathcal{D}_j^{train}, \theta)).$$

3.4 FEDERATED META-LEARNING FOR ACHIEVING REAL-TIME EDGE INTELLIGENCE

As illustrated in Figure 3.1, we consider a platform-aided architecture where a set $\mathcal{S}$ of source edge nodes (each with a task) join force for FedML, and the learned model would be transferred from the platform to a target edge node t (not in $\mathcal{S}$), for rapid adaptation based on its local data. The primary objective of the proposed FedML is to train a meta-model that can quickly adapt to the task at the target edge node to achieve real-time edge intelligence, using only a few local data points. To accomplish this, the meta-model is trained during a meta-learning phase across the source edge nodes in a distributed manner.

3.4.1 PROBLEM FORMULATION

Specifically, we assume that the tasks across edge nodes follow a meta-model, represented by a parametrized function f_θ with parameters $\theta \in \mathbb{R}^d$. For source edge node $i \in \mathcal{S}$, let D_i denote its local dataset $\{(\mathbf{x}_i^1, \mathbf{y}_i^1), \ldots, (\mathbf{x}_i^j, \mathbf{y}_i^j), \ldots, (\mathbf{x}_i^{|D_i|}, \mathbf{y}_i^{|D_i|})\}$, where $|D_i|$ is the dataset size and $(\mathbf{x}^j, \mathbf{y}^j) \in \mathcal{X} \times \mathcal{Y}$ is a sample point with $\mathbf{x}^j$ being the input and $\mathbf{y}^j$ the output. We further assume that $(\mathbf{x}_i^j, \mathbf{y}_i^j)$ follows an unknown distribution P_i. Denote the loss function by $l(\boldsymbol{\theta}, (\mathbf{x}^j, \mathbf{y}^j)) : \mathcal{X} \times \mathcal{Y} \to \mathbb{R}$. The empirical loss function for node i is then defined as

$$L(\boldsymbol{\theta}, D_i) \triangleq \frac{1}{|D_i|} \sum_{\left(\mathbf{x}_i^j, \mathbf{y}_i^j\right) \in D_i} l\left(\boldsymbol{\theta}, \left(\mathbf{x}_i^j, \mathbf{y}_i^j\right)\right), \tag{3.1}$$

which we write as $L_i(\boldsymbol{\theta})$ for brevity. Moreover, we use $L_w(\boldsymbol{\theta})$ to denote the overall loss function across all edge nodes in $\mathcal{S}$:

$$L_w(\boldsymbol{\theta}) \triangleq \sum_{i \in \mathcal{S}} \omega_i L_i(\boldsymbol{\theta}), \tag{3.2}$$

where $\omega_i = \frac{|D_i|}{\sum_{i \in \mathcal{S}} |D_i|}$ and the weight ω_i of each edge node depends on its own local data size.

In the same spirit as MAML, we consider that the target edge node t has K data samples, i.e., $|D_t| = K$. For each source edge node $i \in \mathcal{S}$, D_i is divided into two disjoint sets, the training set D_i^{train} and the testing set D_i^{test}, where $|D_i^{train}| = K$ (assuming $|D_i| > K$ for all $i \in \mathcal{S}$). Given the model parameter $\boldsymbol{\theta}$, the edge node i first updates $\boldsymbol{\theta}$ using one step gradient descent based on D_i^{train}:

$$\boldsymbol{\phi}_i(\boldsymbol{\theta}) = \boldsymbol{\theta} - \alpha \nabla_{\boldsymbol{\theta}} L\left(\boldsymbol{\theta}, D_i^{train}\right), \tag{3.3}$$

with α being the learning rate, and then evaluates the loss $L(\boldsymbol{\phi}_i, D_i^{test})$ for the updated model parameter $\boldsymbol{\phi}_i$ based on D_i^{test}. It follows that the overall objective of the FedML is given by

$$\min_{\boldsymbol{\theta}} \sum_{i \in \mathcal{S}} \omega_i L\left(\boldsymbol{\phi}_i(\boldsymbol{\theta}), D_i^{test}\right). \tag{3.4}$$

Intuitively, by considering how the test error on local testing datasets changes with respect to the updated model parameters, we aim to obtain a model initialization such that small changes in the model parameters, i.e., altering in the direction of the loss gradient, would lead to substantial performance improvements for any task across the edge nodes.

Departing from MAML which assumes that a distribution over tasks is given, we do not require such an assumption here. Instead, we will quantify the node similarity in terms of the variations of the gradients and Hessians of local loss functions with respect to the hyperparameter. Worth noting is that without knowing the data, the platform cannot directly solve the problem (3.4).

3.4.2　FEDERATED META-LEARNING (FedML)

Motivated by federated learning, we propose to solve the meta-learning objective (3.4) in a distributed manner, in the sense that each node locally updates the model parameter $\boldsymbol{\theta}$ based on its own dataset and transmits the updated value to the platform for a global aggregation. To better utilize the local computing resource of edge nodes and reduce the communication cost between the platform and edge nodes which is often a significant bottleneck in wireless networks, each edge node is allowed to locally update $\boldsymbol{\theta}$ for T_0 steps before uploading the results to the platform.

Federated Meta-Training across Source Nodes: More specifically, the platform transfers an initialized $\boldsymbol{\theta}^0$ to all nodes in $\mathcal{S}$ at time $t = 0$. As outlined in Algorithm 3.1, there are two major steps.

- **Local Update**: For $t \neq nT_0$ where $n \in \mathbb{N}^+$, each node $i \in \mathcal{S}$ first updates $\boldsymbol{\theta}_i^t$ using the training dataset D_i^{train} based on (3.3) and then locally updates $\boldsymbol{\theta}_i^t$ again through evaluating $\boldsymbol{\phi}_i^t$ on the testing dataset D_i^{test}:

$$\boldsymbol{\theta}_i^{t+1} = \boldsymbol{\theta}_i^t - \beta \nabla_{\boldsymbol{\theta}} \left(\boldsymbol{\phi}_i^t, D_i^{test} \right), \tag{3.5}$$

where β is the meta learning rate and $\boldsymbol{\theta}_i^{t+1}$ will be used as the starting point for the next iteration at node i.

- **Global Aggregation**: For $t = nT_0$, each node also needs to transmit the updated $\boldsymbol{\theta}_i^{t+1}$ to the platform. The platform then performs a global aggregation to achieve $\boldsymbol{\theta}^{t+1}$:

$$\boldsymbol{\theta}^{t+1} = \sum_{i \in \mathcal{S}} \omega_i \boldsymbol{\theta}_i^{t+1}, \tag{3.6}$$

and sends $\boldsymbol{\theta}^{t+1}$ back to all edge nodes for the next iteration.

The details are summarized in Algorithm 3.1.

Algorithm 3.1 Federated meta-learning (FedML)

    Inputs: M, K, T, T_0, α, β, ω_i for $i \in \mathcal{S}$
    Outputs: Final model parameter $\boldsymbol{\theta}$
1: Platform randomly initializes $\boldsymbol{\theta}^0$ and sends it to all edge nodes in $\mathcal{S}$;
2: **for** $t = 1, 2, ..., T$ **do**
3:     **for** each node $i \in \mathcal{S}$ **do**
4:        Compute the updated parameter with one-step gradient descent using D_i^{train}: $\boldsymbol{\phi}_i^t = \boldsymbol{\theta}_i^t - \alpha \nabla_{\boldsymbol{\theta}} L(\boldsymbol{\theta}_i^t, D_i^{train})$;
5:        Obtain $\boldsymbol{\theta}_i^{t+1}$ based on (3.5) using D_i^{test}; //**local update**
6:        **if** t is a multiple of T_0 **then**
7:           Send $\boldsymbol{\theta}_i^{t+1}$ to the platform;
8:           Receive $\boldsymbol{\theta}^{t+1}$ from the platform where $\boldsymbol{\theta}^{t+1}$ is obtained based on (3.6);
9:           Set $\boldsymbol{\theta}_i^t \leftarrow \boldsymbol{\theta}^{t+1}$; //**global aggregation**
10:       **else**
11:          Set $\boldsymbol{\theta}_i^t \leftarrow \boldsymbol{\theta}_i^{t+1}$;
12: **return** $\boldsymbol{\theta}$.

Fast Adaptation toward Real-time Edge Intelligence at Target Node: Given the model parameter θ from the platform, the target edge node t can quickly adapt the model based on its local dataset D_t and obtain a new model parameter ϕ_t through one step gradient descent:

$$\phi_t = \theta - \alpha \nabla_\theta L(\theta, D_t). \tag{3.7}$$

In a nutshell, instead of training each source edge node to learn a global model as in federated learning, the source nodes join force to learn how to learn quickly with only a few data samples in a distributed manner, i.e., learn θ such that just one step gradient descent from θ can bring up a new model suitable for a specific target node. This greatly improves the fast learning capability at the edge with the collaboration among edge nodes.

3.5   PERFORMANCE ANALYSIS OF FedML

In this section, we seek to answer the following two key questions: (1) What is the convergence performance of the proposed FedML algorithm? (2) Can the fast adaptation at the target node achieve good performance?

3.5.1   CONVERGENCE ANALYSIS

For ease of exposition, we define function $G_i(\theta) \triangleq L_i(\phi_i(\theta))$ and $G(\theta) \triangleq \sum_{i \in S} \omega_i G_i(\theta)$ such that problem (3.4) can be written as:

$$\min_{\theta} \quad G(\theta). \tag{3.8}$$

For convenience, we assume $T = N T_0$ and make the following assumptions to the loss function for all $i \in S$.

Assumption 3.1   Each $L_i(\theta)$ is μ-strongly convex, i.e., for all $\theta, \theta' \in \mathbb{R}^d$,

$$\langle \nabla L_i(\theta) - \nabla L_i(\theta'), \theta - \theta' \rangle \geq \mu \|\theta - \theta'\|^2.$$

Assumption 3.2   Each $L_i(\theta)$ is H-smooth, i.e., for all $\theta, \theta' \in \mathbb{R}^d$,

$$\|\nabla L_i(\theta) - \nabla L_i(\theta')\| \leq H \|\theta - \theta'\|,$$

and there exist constant B such that for all $\theta \in \mathbb{R}^d$

$$\|\nabla L_i(\theta)\| \leq B.$$

Assumption 3.3   The Hessian of each $L_i(\boldsymbol{\theta})$ is ρ-Lipschitz, i.e., for all $\boldsymbol{\theta}, \boldsymbol{\theta}' \in \mathbb{R}^d$,

$$\|\nabla^2 L_i(\boldsymbol{\theta}) - \nabla^2 L_i(\boldsymbol{\theta}')\| \leq \rho\|\boldsymbol{\theta} - \boldsymbol{\theta}'\|.$$

Assumption 3.4   There exists constants δ_i and σ_i such that for all $\boldsymbol{\theta} \in \mathbb{R}^d$

$$\|\nabla L_i(\boldsymbol{\theta}) - \nabla L_w(\boldsymbol{\theta})\| \leq \delta_i \quad \text{and} \quad \left\|\nabla^2 L_i(\boldsymbol{\theta}) - \nabla^2 L_w(\boldsymbol{\theta})\right\| \leq \sigma_i.$$

Assumptions 3.1–3.2 are standard and hold in many machine learning applications, e.g., in logistic regression over a bounded domain and squared-SVM. Assumption 3.3 is concerned with the high-order smoothness of the local loss function at each edge node, which makes it possible to characterize the landscape of the local meta-learning objective function. Assumption 3.4 is imposed to capture the node similarity. Specifically, we impose the condition that the variations of the gradients and Hessians of local loss functions (with respect to the hyper-parameter) across edge nodes are upper bounded by some constant. Intuitively, a small (large) constant indicates that the tasks are more (less) similar, and this constant can be tuned over a wide range to obtain a general understanding. Further, for a task distribution with task gradients uniformly bounded above, Assumption 3.4 follows directly because $\|\nabla L_i(\boldsymbol{\theta}) - \nabla L_w(\boldsymbol{\theta})\|$ (also Hessian) can be viewed as the distance between a typical realization and the sample average. In a nutshell, the task similarity assumption here is more general and realistic than the assumption in meta-learning that all tasks follow a (known) distribution. It is worth noting that these assumptions do not trivialize the meta-learning setting.

To characterize the convergence behavior of the FedML algorithm, we first examine the structural properties of the global meta-learning objective $\mathbf{G}(\boldsymbol{\theta})$. Next, we study the impact of the task similarity across different edge nodes on the convergence performance of the FedML, which is further complicated by multiple local updates at each node to reduce the communication overhead.

Convexity and Smoothness of the Federated Meta-Learning Objective Function: Based on Theorem 1 in Finn et al. [2019], we first have the following result about the structural properties of function $G(\boldsymbol{\theta})$.

Lemma 3.5   Suppose Assumptions 3.1–3.3 hold. When $\alpha \leq \min\{\frac{\mu}{2\mu H + \rho B}, \frac{1}{\mu}\}$, $G(\boldsymbol{\theta})$ is μ'-strongly convex and H'-smooth, where $\mu' = \mu(1 - \alpha H)^2 - \alpha\rho B$ and $H' = H(1 - \alpha\mu)^2 + \alpha\rho B$.

Proof. We first show that $G_i(\theta)$ is μ'-strongly convex and L'-smooth. Specifically, observe that

$$\|\nabla G_i(\theta) - \nabla G_i(\theta')\|$$
$$=\|\nabla L_i(\phi_i) - \alpha\nabla^2 L_i(\theta)\nabla L_i(\phi_i) - \nabla L_i(\phi'_i) + \alpha\nabla^2 L_i(\theta')\nabla L_i(\phi'_i)\|$$
$$=\|\nabla L_i(\phi_i) - \alpha\nabla^2 L_i(\theta)\nabla L_i(\phi_i) + \alpha\nabla^2 L_i(\theta')\nabla L_i(\phi'_i) - \nabla L_i(\phi'_i)$$
$$+ \alpha\nabla^2 L_i(\theta)\nabla L_i(\phi'_i) - \alpha\nabla^2 L_i(\theta)\nabla L_i(\phi'_i)\|$$
$$=\|[I - \alpha\nabla^2 L_i(\theta)][\nabla L_i(\phi_i) - \nabla L_i(\phi'_i)] - \alpha\nabla L_i(\phi'_i)[\nabla^2 L_i(\theta) - \nabla^2 L_i(\theta')]\|, \tag{3.9}$$

where $\phi_i = \theta - \alpha\nabla L_i(\theta)$ and $\phi'_i = \theta' - \alpha\nabla L_i(\theta')$.

To establish the convexity, it suffices to show $\|\nabla G_i(\theta) - \nabla G_i(\theta')\| \geq \mu'\|\theta - \theta'\|$. It can be seen from (3.9) that

$$\|\nabla G_i(\theta) - \nabla G_i(\theta')\| \geq (1 - \alpha H)\|\nabla L_i(\phi_i) - \nabla L_i(\phi'_i)\|$$
$$- \alpha\|\nabla L_i(\phi'_i)\|\|\nabla^2 L_i(\theta) - \nabla^2 L_i(\theta')\|$$
$$\geq \mu(1 - \alpha H)\|\phi_i - \phi'_i\| - \alpha\rho B\|\theta - \theta'\|. \tag{3.10}$$

Since $\nabla\phi_i = I - \alpha\nabla^2 L_i(\theta)$, it follows from Assumption 1 and 2 that $1 - \alpha H \leq \nabla\phi_i \leq 1 - \alpha\mu$, which indicates

$$(1 - \alpha H)\|\theta - \theta'\| \leq \|\phi_i - \phi'_i\| \leq (1 - \alpha\mu)\|\theta - \theta'\|. \tag{3.11}$$

Combining (3.10) and (3.11), we have

$$\|\nabla G_i(\theta) - \nabla G_i(\theta')\| \geq \mu'\|\theta - \theta'\|,$$

where $\mu' - \mu(1 - \alpha H)^2 - \alpha\rho B > 0$.

To establish the smoothness, it suffices to show $\|\nabla G_i(\theta) - \nabla G_i(\theta')\| \leq H'\|\theta - \theta'\|$. From (3.9) and (3.11), we have

$$\|\nabla G_i(\theta) - \nabla G_i(\theta')\| \leq (1 - \alpha\mu)\|\nabla L_i(\phi_i) - \nabla L_i(\phi'_i)\|$$
$$+ \alpha\|\nabla L_i(\phi'_i)\|\|\nabla^2 L_i(\theta) - \nabla^2 L_i(\theta')\|$$
$$\leq H(1 - \alpha\mu)\|\phi_i - \phi'_i\| + \alpha\rho B\|\theta - \theta'\|$$
$$\leq H'\|\theta - \theta'\|,$$

where $H' = H(1 - \alpha\mu)^2 + \alpha\rho B$, thereby completing the proof of Lemma 3.5. $\qquad\square$

Lemma 3.5 indicates that when the learning rate α is relatively small, the meta-learning objective function $G(\theta)$ formed by the one-step gradient descent on local datasets is as well-behaved as the local loss function.

Bounded Dissimilarity across Local Learning tasks: Next, we characterizes the impact of the similarity across local learning tasks.

Theorem 3.6 *Suppose Assumptions 3.2 and 3.4 hold. Then there exists a constant C such that*

$$\|\nabla G_i(\theta) - \nabla G(\theta)\| \leq \delta_i + \alpha C(H\delta_i + B\sigma_i + \tau),$$

where $\tau = \sum_{i \in S} \omega_i \delta_i \sigma_i$.

Proof. Observe that $\nabla G_i(\theta) = \nabla L_i(\phi_i) - \alpha \nabla^2 L_i(\theta) \nabla L_i(\phi_i)$ involves the product between Hessian matrix and gradient, which admits an upper bound outlined as follows:

$$\|\nabla^2 L_i(\theta)\nabla L_i(\theta) - \sum_{i \in \mathcal{S}} \omega_i \nabla^2 L_i(\theta)\nabla L_i(\theta)\|$$

$$=\|\nabla^2 L_i(\theta)\nabla L_i(\theta) - \nabla^2 L_w(\theta)\nabla L_w(\theta) + \nabla^2 L_w(\theta)\nabla L_w(\theta) - \sum_{i \in \mathcal{S}} \omega_i \nabla^2 L_i(\theta)\nabla L_i(\theta)\|$$

$$\leq\|\nabla^2 L_i(\theta)\nabla L_i(\theta) - \nabla^2 L_i(\theta)\nabla L_w(\theta)\| + \|\nabla^2 L_i(\theta)\nabla L_w(\theta) - \nabla^2 L_w(\theta)\nabla L_w(\theta\|$$

$$+ \|\sum_{i \in \mathcal{S}} \omega_i [(\nabla L_i(\theta) - \nabla L_w(\theta))(\nabla^2 L_i(\theta) - \nabla^2 L_w(\theta))]\|$$

$$\leq\|\nabla^2 L_i(\theta)\|\|\nabla L_i(\theta) - \nabla L_w(\theta)\| + \|\nabla L_w(\theta)\|\|\nabla^2 L_i(\theta) - \nabla^2 L_w(\theta)\|$$

$$+ \sum_{i \in \mathcal{S}} \omega_i \|\nabla L_i(\theta) - \nabla L_w(\theta)\|\|\nabla^2 L_i(\theta) - \nabla^2 L_w(\theta)\|$$

$$\leq H\delta_i + B\sigma_i + \tau. \tag{3.12}$$

Next, it follows from Taylor's Theorem that

$$\nabla L_i(\phi_i) = \nabla L_i(\theta) + \nabla^2 L_i(\theta)(\phi_i - \theta) + O(\|\phi_i - \theta\|^2).$$

That is to say,

$$\nabla L_i(\phi_i) = \nabla L_i(\theta) - \alpha \nabla^2 L_i(\theta)\nabla L_i(\theta) + O(\alpha^2 B^2).$$

Therefore, we have

$$\|\nabla G_i(\theta) - \nabla G(\theta)\|$$

$$=\|[I - \alpha\nabla^2 L_i(\theta)]\nabla L_i(\phi_i) - \sum_{i \in \mathcal{S}} \omega_i [I - \alpha\nabla^2 L_i(\theta)]\nabla L_i(\phi_i)\|$$

$$=\|[I - \alpha\nabla^2 L_i(\theta)][\nabla L_i(\phi_i) - \nabla L_i(\theta) + \nabla L_i(\theta)]$$

$$- \sum_{i \in \mathcal{S}} \omega_i [I - \alpha\nabla^2 L_i(\theta)][\nabla L_i(\phi_i)\nabla - L_i(\theta) + \nabla L_i(\theta)]\|$$

$$=\|\nabla L_i(\theta) - \nabla L_w(\theta) - 2\alpha\nabla^2 L_i(\theta)\nabla L_i(\theta) + 2\alpha \sum_{i \in \mathcal{S}} \omega_i \nabla^2 L_i(\theta)\nabla L_i(\theta) + O(\alpha^2 B^2)$$

$$+ \alpha^2 [\nabla^2 L_i(\theta)]^2 \nabla L_i(\theta) - \alpha^2 \sum_{i \in \mathcal{S}} \omega_i [\nabla^2 L_i(\theta)]^2 \nabla L_i(\theta)\|$$

$$\leq\|\nabla L_i(\theta) - \nabla L_w(\theta)\| + O(\alpha^2 B^2) + 2\alpha\|\nabla^2 L_i(\theta)\nabla L_i(\theta) - \sum_{i \in \mathcal{S}} \omega_i \nabla^2 L_i(\theta)\nabla L_i(\theta)\|$$

$$+ \alpha^2 \|[\nabla^2 L_i(\theta)]^2\nabla L_i(\theta) - \sum_{i \in \mathcal{S}} \omega_i [\nabla^2 L_i(\theta)]^2\nabla L_i(\theta)\|$$

$$\leq\delta_i + 2\alpha(H\delta_i + B\sigma_i + \tau) + O(\alpha^2 B^2) + \alpha^2 \|[\nabla^2 L_i(\theta)]^2\nabla L_i(\theta) - \alpha^2 \sum_{i \in \mathcal{S}} \omega_i [\nabla^2 L_i(\theta)]^2\nabla L_i(\theta)\|.$$

Along the same line as in finding an upper bound for $\|\nabla^2 L_i(\theta)\nabla L_i(\theta) - \sum_{i \in \mathcal{S}} \omega_i \nabla^2 L_i(\theta)\nabla L_i(\theta)\|$, we can find an upper bound on the last term of the above inequality with $\alpha^2(H\delta_i' + B\sigma_i + \tau')$ where $\delta_i' = H\delta_i + B\sigma_i + \tau$ and $\tau' = \sum_{i \in \mathcal{S}} \omega_i \delta_i' \sigma_i$. We conclude that when α is suitably small, there exists a constant C such that the right hand side of the above inequality is upper bounded by $\delta_i + \alpha C(H\delta_i + B\sigma_i + \tau)$. $\square$

Given the bounded variance of gradients and Hessians of local loss functions, we can find upper bounds on the gradient variance of the local meta-learning objective functions, while still preserving the node heterogeneity. As a sanity check, if all the edge nodes have same data points, it follows that $\delta_i = \sigma_i = 0$ for all $i \in \mathcal{S}$. Consequently, all edge nodes have the same local learning objective.

Based on Lemma 3.5 and Theorem 3.6, we can have the following result about the convergence performance of the FedML algorithm.

Theorem 3.7   *Suppose that Assumptions 3.1–3.4 hold, and the learning rates α and β are chosen to satisfy that $\alpha \leq \min\{\frac{\mu}{2\mu H + \rho B}, \frac{1}{\mu}\}$ and $\beta < \min\{\frac{1}{2\mu'}, \frac{2}{H'}\}$. Let $\delta = \sum_{i \in \mathcal{S}} \omega_i \delta_i$ and $\sigma = \sum_{i \in \mathcal{S}} \omega_i \sigma_i$. Then*

$$G(\boldsymbol{\theta}^T) - G(\boldsymbol{\theta}^\star) \leq \xi^T [G(\boldsymbol{\theta}^0) - G(\boldsymbol{\theta}^\star)] + \frac{B(1 - \alpha\mu)}{1 - \xi^{T_0}} h(T_0),$$

where $\xi = 1 - 2\beta\mu'\left(1 - \frac{H'\beta}{2}\right)$, $h(x) \triangleq \frac{\alpha'}{\beta H'}[(1 + \beta H')^x - 1] - \alpha' x$, $\alpha' = \beta[\delta + \alpha C(H\delta + B\sigma + \tau)]$.

Proof. Following the same method as in Wang et al. [2019a], we first define a virtual sequence for global aggregation at each iteration as $\boldsymbol{v}^t_{[n]}$ for $t \in [(n-1)T_0, nT_0]$, where the interval $[(n-1)T_0, nT_0]$ is denoted as $[n]$. More specifically,

$$\boldsymbol{v}^{t+1}_{[n]} = \boldsymbol{v}^t_{[n]} - \beta\nabla G\left(\boldsymbol{v}^t_{[n]}\right), \tag{3.13}$$

and $\boldsymbol{v}^t_{[n]}$ is assumed to be "synchronized" with $\boldsymbol{\theta}^t$ at the beginning of interval $[n]$, i.e., $\boldsymbol{v}^{(n-1)T_0}_{[n]} = \boldsymbol{\theta}^{(n-1)T_0}$, where $\boldsymbol{\theta}^{(n-1)T_0}$ is the weighted average of local parameters $\boldsymbol{\theta}^{(n-1)T_0}_i$ as shown in (3.6). To show the convergence, we first analyze the gap between the virtual global parameter $\boldsymbol{v}^t_{[n]}$ and the local weighted average $\boldsymbol{\theta}^t$ during each interval, and then evaluate the convergence performance of $\boldsymbol{\theta}^t$ through evaluating the convergence performance of virtual sequence $\boldsymbol{v}^t_{[n]}$ by taking the gap into consideration.

To analyze the gap between $\boldsymbol{v}^t_{[n]}$ and $\boldsymbol{\theta}^t$ during interval $[n]$, we first look into the gap between $\boldsymbol{v}^t_{[n]}$ and the local update $\boldsymbol{\theta}^t_i$. Specifically,

$$\begin{aligned}
\|\boldsymbol{\theta}^{t+1}_i - \boldsymbol{v}^{t+1}_{[n]}\| &= \|\boldsymbol{\theta}^t_i - \beta\nabla G_i(\boldsymbol{\theta}^t_i) - \boldsymbol{v}^t_{[n]} + \beta\nabla G(\boldsymbol{v}^t_{[n]})\| \\
&\leq \|\boldsymbol{\theta}^t_i - \boldsymbol{v}^t_{[n]}\| + \beta\|\nabla G_i(\boldsymbol{\theta}^t_i) - \nabla G(\boldsymbol{v}^t_{[n]})\| \\
&\leq \|\boldsymbol{\theta}^t_i - \boldsymbol{v}^t_{[n]}\| + \beta\|\nabla G_i(\boldsymbol{\theta}^t_i) - \nabla G_i(\boldsymbol{v}^t_{[n]})\| + \beta\|\nabla G_i(\boldsymbol{v}^t_{[n]}) - \nabla G(\boldsymbol{v}^t_{[n]})\| \\
&\leq (1 + \beta H')\|\boldsymbol{\theta}^t_i - \boldsymbol{v}^t_{[n]}\| + \beta[\delta_i + \alpha C(H\delta_i + B\sigma_i + \tau)]. \tag{3.14}
\end{aligned}$$

By induction, we can show that $\|\boldsymbol{\theta}_i^t - \boldsymbol{v}_{[n]}^t\| \le g(t - (n-1)T_0)$ where $g(x) \triangleq \frac{\delta_i + \alpha C(H\delta_i + B\sigma_i + \tau)}{H'}[(1 + \beta H')^x - 1]$. Therefore, for $t \in [(n-1)T_0, nT_0)$ we can get

$$\begin{aligned}
\|\boldsymbol{\theta}^{t+1} - \boldsymbol{v}_{[n]}^{t+1}\| &= \|\sum_{i \in S} \omega_i \boldsymbol{\theta}_i^{t+1} - \boldsymbol{v}_{[n]}^{t+1}\| \\
&= \|\boldsymbol{\theta}^t - \beta \sum_{i \in S} \omega_i \nabla G_i(\boldsymbol{\theta}_i^t) - \boldsymbol{v}_{[n]}^t + \beta \nabla G(\boldsymbol{v}_{[n]}^t)\| \\
&\le \|\boldsymbol{\theta}^t - \boldsymbol{v}_{[n]}^t\| + \beta \|\sum_{i \in S} \omega_i (\nabla G_i(\boldsymbol{\theta}_i^t) - \nabla G_i(\boldsymbol{v}_{[n]}^t))\| \\
&\le \|\boldsymbol{\theta}^t - \boldsymbol{v}_{[n]}^t\| + \beta \sum_{i \in S} \omega_i \|\nabla G_i(\boldsymbol{\theta}_i^t) - \nabla G_i(\boldsymbol{v}_{[n]}^t)\| \\
&\le \|\boldsymbol{\theta}^t - \boldsymbol{v}_{[n]}^t\| + \beta H' \sum_{i \in S} \omega_i \|\boldsymbol{\theta}_i^t - \boldsymbol{v}_{[n]}^t\| \\
&\le \|\boldsymbol{\theta}^t - \boldsymbol{v}_{[n]}^t\| + \beta H' \sum_{i \in S} \omega_i g(t - (n-1)T_0) \\
&= \|\boldsymbol{\theta}^t - \boldsymbol{v}_{[n]}^t\| + \alpha'[(1 + \beta H')^{t-(n-1)T_0} - 1],
\end{aligned} \tag{3.15}$$

where $\alpha' = \beta[\delta + \alpha C(H\delta + B\sigma + \tau)]$. Since $\boldsymbol{\theta}^{(n-1)T_0} = \boldsymbol{v}_{[n]}^{(n-1)T_0}$, we have

$$\begin{aligned}
\|\boldsymbol{\theta}^t - \boldsymbol{v}_{[n]}^t\| &\le \sum_{j=1}^{t-(n-1)T_0} \{\alpha'[(1 + \beta H')^j - 1]\} \\
&= \frac{\alpha'}{\beta H'}[(1 + \beta H')^{t-(n-1)T_0} - 1] - \alpha'[t - (n-1)T_0] \\
&\triangleq h(t - (n-1)T_0).
\end{aligned} \tag{3.16}$$

Next, we evaluate the convergence performance of virtual sequence $\boldsymbol{v}_{[n]}^t$ during the interval $[n]$ for $t \in [(n-1)T_0, nT_0]$. Since $G(\cdot)$ is H'-smooth, we can have

$$\begin{aligned}
G(\boldsymbol{v}_{[n]}^{t+1}) - G(\boldsymbol{v}_{[n]}^t) &\le \langle \nabla G(\boldsymbol{v}_{[n]}^t), \boldsymbol{v}_{[n]}^{t+1} - \boldsymbol{v}_{[n]}^t \rangle + \frac{H'}{2}\|\boldsymbol{v}_{[n]}^{t+1} - \boldsymbol{v}_{[n]}^t\|^2 \\
&\le -\beta \left(1 - \frac{H'\beta}{2}\right)\|\nabla G(\boldsymbol{v}_{[n]}^t)\|^2.
\end{aligned} \tag{3.17}$$

Moreover, since $G(\cdot)$ is μ'-strongly convex, it follows that

$$G(\boldsymbol{v}_{[n]}^t) \le G(\boldsymbol{\theta}^\star) + \frac{1}{2\mu'}\|\nabla G(\boldsymbol{v}_{[n]}^t)\|^2. \tag{3.18}$$

Combining (3.17) and (3.18) gives us

$$G(\boldsymbol{v}_{[n]}^{t+1}) - G(\boldsymbol{v}_{[n]}^t) \le -2\beta\mu' \left(1 - \frac{H'\beta}{2}\right)[G(\boldsymbol{v}_{[n]}^t) - G(\boldsymbol{\theta}^\star)]$$

which is equivalent with

$$G(\boldsymbol{v}_{[n]}^{t+1}) - G(\boldsymbol{\theta}^\star) \le \left[1 - 2\beta\mu' \left(1 - \frac{H'\beta}{2}\right)\right][G(\boldsymbol{v}_{[n]}^t) - G(\boldsymbol{\theta}^\star)] = \xi[G(\boldsymbol{v}_{[n]}^t) - G(\boldsymbol{\theta}^\star)],$$

where $\xi = 1 - 2\beta\mu'\left(1 - \frac{H'\beta}{2}\right) \in (0,1)$ given $\beta < \min\{\frac{1}{2\mu'}, \frac{2}{H'}\}$. Iteratively, we can obtain

$$
\begin{aligned}
G(v_{[n]}^{nT_0}) - G(\theta^\star) &\leq \xi[G(v_{[n]}^{nT_0-1}) - G(\theta^\star)] \\
&\leq \xi^2[G(v_{[n]}^{nT_0-2}) - G(\theta^\star)] \\
&\cdots \\
&\leq \xi^{T_0}[G(v_{[n]}^{(n-1)T_0}) - G(\theta^\star)] \\
&= \xi^{T_0}[G(v_{[n-1]}^{(n-1)T_0}) - G(\theta^\star)] + \xi^{T_0}[G(v_{[n]}^{(n-1)T_0}) - G(v_{[n-1]}^{(n-1)T_0})].
\end{aligned}
\tag{3.19}
$$

Note that

$$
\begin{aligned}
\|\nabla G(\theta)\| &= \|\sum_{i \in \mathcal{S}} \omega_i[(I - \alpha\nabla^2 L_i(\theta))\nabla L_i(\phi_i)]\| \\
&\leq \sum_{i \in \mathcal{S}} \omega_i\|I - \alpha\nabla^2 L_i(\theta)\|\|\nabla L_i(\phi_i)\| \\
&\leq (1 - \alpha\mu)B.
\end{aligned}
\tag{3.20}
$$

From the Mean Value Theorem, we conclude that $\|G(\theta) - G(\theta')\| \leq (1 - \alpha\mu)B\|\theta - \theta'\|$. Hence, we can upper bound $G(v_{[n]}^{(n-1)T_0}) - G(v_{[n-1]}^{(n-1)T_0})$ as follows:

$$
\begin{aligned}
G(v_{[n]}^{(n-1)T_0}) - G(v_{[n-1]}^{(n-1)T_0}) &= G(\theta^{(n-1)T_0}) - G(v_{[n-1]}^{(n-1)T_0}) \\
&\leq B(1 - \alpha\mu)\|\theta^{(n-1)T_0} - v_{[n-1]}^{(n-1)T_0}\| \\
&\overset{(a)}{\leq} B(1 - \alpha\mu)h(T_0),
\end{aligned}
\tag{3.21}
$$

where (a) is from (3.16). Substitute (3.21) in (3.19), we have

$$
G(v_{[n]}^{nT_0}) - G(\theta^\star) \leq \xi^{T_0}[G(v_{[n-1]}^{(n-1)T_0}) - G(\theta^\star)] + \xi^{T_0}B(1 - \alpha\mu)h(T_0).
$$

Iteratively, it follows that

$$
\begin{aligned}
G(v_{[N]}^{NT_0}) - G(\theta^\star) &\leq \xi^{T_0}[G(v_{[N-1]}^{(N-1)T_0}) - G(\theta^\star)] + \xi^{T_0}B(1 - \alpha\mu)h(T_0) \\
&\leq \xi^{2T_0}[G(v_{[N-2]}^{(N-2)T_0}) - G(\theta^\star)] + (\xi^{T_0} + \xi^{2T_0})B(1 - \alpha\mu)h(T_0) \\
&\cdots \\
&\leq \xi^{(N-1)T_0}[G(v_{[1]}^{T_0}) - G(\theta^\star)] + \sum_{j=1}^{N-1}\xi^{jT_0}B(1 - \alpha\mu)h(T_0) \\
&\leq \xi^{NT_0}[G(v_{[1]}^0) - G(\theta^\star)] + \sum_{j=1}^{N-1}\xi^{jT_0}B(1 - \alpha\mu)h(T_0) \\
&= \xi^{NT_0}[G(\theta^0) - G(\theta^\star)] + \sum_{j=1}^{N-1}\xi^{jT_0}B(1 - \alpha\mu)h(T_0).
\end{aligned}
$$

Therefore, we can conclude that

$$
\begin{aligned}
G(\boldsymbol{\theta}^T) - G(\boldsymbol{\theta}^\star) &= G(v_{[N+1]}^{NT_0}) - G(\boldsymbol{\theta}^\star) \\
&= G(v_{[N]}^{NT_0}) - G(\boldsymbol{\theta}^\star) + G(v_{[N+1]}^{NT_0}) - G(v_{[N]}^{NT_0}) \\
&\leq \xi^{NT_0}[G(\boldsymbol{\theta}^0) - G(\boldsymbol{\theta}^\star)] + \sum_{j=1}^{N-1} \xi^{jT_0} B(1-\alpha\mu)h(T_0) + B(1-\alpha\mu)h(T_0) \\
&= \xi^T[G(\boldsymbol{\theta}^0) - G(\boldsymbol{\theta}^\star)] + \frac{1-\xi^T}{1-\xi^{T_0}} B(1-\alpha\mu)h(T_0) \\
&\leq \xi^T[G(\boldsymbol{\theta}^0) - G(\boldsymbol{\theta}^\star)] + \frac{B(1-\alpha\mu)}{1-\xi^{T_0}}h(T_0),
\end{aligned}
$$

thereby completing the proof of Theorem 3.7. $\square$

Intuitively, the term $\frac{B(1-\alpha\mu)}{1-\xi^{T_0}}h(T_0)$ captures the error introduced by both task dissimilarity and multiple local updates through the function $h(T_0)$. Specifically, observe that $h(T_0)$ increases with δ and σ, which clearly indicates how the task similarity and the number of local update steps T_0 impact the convergence performance, i.e., given a fixed duration T the convergence error decreases with the task similarity while increasing with the number of local update steps when T_0 is large. Correspondingly, the platform is able to balance between the platform-edge communication cost and the local computation cost via controlling the number of local update steps T_0 per communication round, depending on the task similarity across the edge nodes.

Different from MAML, multiple local updates are allowed in Algorithm 3.1 to reduce the communication cost, which has nontrivial impact on the convergence behavior. As shown in Theorem 3.7, the convergence gap for FedML would be large if the number of local update steps T_0 is large even if the tasks are very similar. When $T_0 = 1$, i.e., each edge node only updates the model locally for one iteration, the term $\frac{B(1-\alpha\mu)}{1-\xi^{T_0}}h(T_0)$ disappears because $h(1) = 0$. We have the following result for this case.

Corollary 3.8   *Suppose that Assumptions 3.1–3.4 hold, and the learning rates α and β are chosen to satisfy that $\alpha \leq \min\{\frac{\mu}{2\mu H+\rho B}, \frac{1}{\mu}\}$ and $\beta < \min\{\frac{1}{2\mu'}, \frac{2}{H'}\}$. When $T_0 = 1$, $G(\boldsymbol{\theta}^T) - G(\boldsymbol{\theta}^\star) \leq \xi^T[G(\boldsymbol{\theta}^0) - G(\boldsymbol{\theta}^\star)]$.*

3.5.2   PERFORMANCE EVALUATION OF FAST ADAPTATION

The fast learning performance at the target edge node t depends on not only its local sample size D_t but also the similarity with the source edge nodes in the FedML. Denote $\boldsymbol{\theta}_c$ as the output of the federated meta-learning at the platform and $\boldsymbol{\theta}_c^\star$ as the optimal meta-learning model. We assume that the convergence error $\|\boldsymbol{\theta}_c - \boldsymbol{\theta}_c^\star\|$ of the FedML algorithm is upper bounded by ϵ_c. For convenience, we further define $L_t^\star(\boldsymbol{\theta})$ as the local average loss over the underlying data distribution P_t:

$$
L_t^\star(\boldsymbol{\theta}) \triangleq \mathbb{E}_{(\mathbf{x}_t^j, \mathbf{y}_t^j) \sim P_t} l(\boldsymbol{\theta}, (\mathbf{x}_t^j, \mathbf{y}_t^j)). \tag{3.22}
$$

Then, the empirical loss $L_t(\boldsymbol{\theta})$ is the sample average approximation of $L_t^\star(\boldsymbol{\theta})$. Let $\boldsymbol{\phi}_t = \boldsymbol{\theta}_c - \alpha \nabla L_t(\boldsymbol{\theta}_c)$ and $\boldsymbol{\phi}_t^\star = \arg\min L_t^\star(\boldsymbol{\phi}) = \boldsymbol{\theta}_t^\star - \alpha \nabla L_t^\star(\boldsymbol{\theta}_t^\star)$. The following result characterizes the trade-off between the target-source similarity and local sample size.

Theorem 3.9   *Suppose $l(\boldsymbol{\theta}, (\mathbf{x}_t^j, \mathbf{y}_t^j))$ is H-smooth with respect to $\boldsymbol{\theta}$ for all $\boldsymbol{\theta} \in \mathbb{R}^d$ and $(\mathbf{x}_t^j, \mathbf{y}_t^j) \sim P_t$. For any $\epsilon > 0$, there exist positive constants C_t and $\eta = \eta(\epsilon)$ such that with probability at least $1 - C_t e^{-\tilde{K}\eta}$ we can have*

$$\|L_t^\star(\boldsymbol{\phi}_t) - L_t^\star(\boldsymbol{\phi}_t^\star)\| \leq \alpha H \epsilon + H(1 + \alpha H)\epsilon_c + H(1 + \alpha H)\|\boldsymbol{\theta}_t^\star - \boldsymbol{\theta}_c^\star\|.$$

Proof. Recall that the optimal model parameter is denoted as $\boldsymbol{\phi}_t^\star = \boldsymbol{\theta}_t^\star - \alpha \nabla L_t^\star(\boldsymbol{\theta}_t^\star)$, i.e., $\boldsymbol{\phi}_t^\star$ can be obtained through one-step gradient update from parameter $\boldsymbol{\theta}_t^\star$. However, through the fast adaptation from the meta-learned model $\boldsymbol{\theta}_c$, we have $\boldsymbol{\phi}_t = \boldsymbol{\theta}_c - \alpha \nabla L_t(\boldsymbol{\theta}_c)$, where $\boldsymbol{\theta}_c$ can be regarded as an estimation of $\boldsymbol{\theta}_t^\star$ and $L_t(\cdot)$ is the sample average approximation of $L_t^\star(\cdot)$. To evaluate the learning performance at the target, we next evaluate the gap between $L_t^\star(\boldsymbol{\phi}_t^\star)$ and $L_t^\star(\boldsymbol{\phi}_t)$, i.e., the gap between the optimal loss and the actual loss based on learned parameter $\boldsymbol{\phi}_t$.

For convenience, denote $\tilde{\boldsymbol{\phi}}_t = \boldsymbol{\theta}_t^\star - \alpha \nabla L_t(\boldsymbol{\theta}_t^\star)$. Note that

$$\|\boldsymbol{\phi}_t - \boldsymbol{\phi}_t^\star\| = \|\boldsymbol{\phi}_t - \tilde{\boldsymbol{\phi}}_t + \tilde{\boldsymbol{\phi}}_t - \boldsymbol{\phi}_t^\star\| \leq \underbrace{\|\boldsymbol{\phi}_t - \tilde{\boldsymbol{\phi}}_t\|}_{(a)} + \underbrace{\|\tilde{\boldsymbol{\phi}}_t - \boldsymbol{\phi}_t^\star\|}_{(b)}, \tag{3.23}$$

where (a) represents the error introduced by the gap between meta-learned model $\boldsymbol{\theta}_c$ and the target optimal $\boldsymbol{\theta}_t^\star$, and (b) captures the error from the sample average approximation of the loss function.

We have the following bound on the term in (a):

$$\begin{aligned}
\|\boldsymbol{\phi}_t - \tilde{\boldsymbol{\phi}}_t\| &= \|\boldsymbol{\theta}_c - \boldsymbol{\theta}_t^\star - \alpha(\nabla L_t(\boldsymbol{\theta}_c) - \nabla L_t(\boldsymbol{\theta}_t^\star))\| \\
&\leq \|\boldsymbol{\theta}_c - \boldsymbol{\theta}_t^\star\| + \alpha\|\nabla L_t(\boldsymbol{\theta}_c) - \nabla L_t(\boldsymbol{\theta}_t^\star)\| \\
&\leq (1 + \alpha H)\|\boldsymbol{\theta}_c - \boldsymbol{\theta}_t^\star\| \\
&= (1 + \alpha H)\|\boldsymbol{\theta}_c - \boldsymbol{\theta}_c^\star + \boldsymbol{\theta}_c^\star - \boldsymbol{\theta}_t^\star\| \\
&\leq (1 + \alpha H)[\|\boldsymbol{\theta}_c - \boldsymbol{\theta}_c^\star\| + \|\boldsymbol{\theta}_c^\star - \boldsymbol{\theta}_t^\star\|] \\
&\leq (1 + \alpha H)[\epsilon_c + \|\boldsymbol{\theta}_c^\star - \boldsymbol{\theta}_t^\star\|]. \tag{3.24}
\end{aligned}$$

To evaluate the term in (b), we first note that $\|\tilde{\boldsymbol{\phi}}_t - \boldsymbol{\phi}_t^\star\| = \alpha\|\nabla L_t(\boldsymbol{\theta}_t^\star) - \nabla L_t^\star(\boldsymbol{\theta}_t^\star)\|$. Here, $\nabla L_t(\cdot) = \frac{1}{K}\sum_{(\mathbf{x}_t^j, \mathbf{y}_t^j) \in D_t} \nabla l(\cdot, (\mathbf{x}_t^j, \mathbf{y}_t^j))$, and $\nabla L_t^\star(\cdot) = \mathbb{E}_{(\mathbf{x}_t, \mathbf{y}_t) \sim P_t} \nabla l(\cdot, (\mathbf{x}_t, \mathbf{y}_t))$. Define $q_t(\cdot) \triangleq \nabla l_t(\cdot)$. Then

$$Q_t(\boldsymbol{\theta}_t^\star) \triangleq \frac{1}{K} \sum_{(\mathbf{x}_t^j, \mathbf{y}_t^j) \in D_t} q_t(\boldsymbol{\theta}_t^\star, (\mathbf{x}_t^j, \mathbf{y}_t^j)) = \nabla L_t(\boldsymbol{\theta}_t^\star),$$

and

$$Q_t^\star(\boldsymbol{\theta}_t^\star) \triangleq \mathbb{E}_{(\mathbf{x}_t,\mathbf{y}_t)\sim P_t} q_t(\boldsymbol{\theta}_t^\star, (\mathbf{x}_t, \mathbf{y}_t)) = \nabla L_t^\star(\boldsymbol{\theta}_t^\star).$$

Clearly, $Q_t(\cdot)$ is the sample average approximation of $Q_t^\star(\cdot)$. Since $l(\boldsymbol{\theta}, (\mathbf{x}_t^j, \mathbf{y}_t^j))$ is H-smooth for all $\boldsymbol{\theta} \in \mathbb{R}^d$ and $(\mathbf{x}_t^j, \mathbf{y}_t^j) \sim P_t$, from Theorem 7.73 in Shapiro et al. [2009] about uniform law of large numbers, we can know that for any $\epsilon > 0$ there exist positive constants C_t and $\eta = \eta(\epsilon)$ such that

$$Pr\left\{\sup_{\boldsymbol{\theta}\in\Theta} \|Q_t(\boldsymbol{\theta}) - Q_t^\star(\boldsymbol{\theta})\| \geq \epsilon\right\} \leq C_t e^{-K\eta},$$

where $\Theta = \{\boldsymbol{\theta} \mid \|\boldsymbol{\theta} - \boldsymbol{\theta}_c\| \leq D\}$ and D is some constant. Hence, we have

$$Pr\left\{\|\tilde{\boldsymbol{\phi}}_t - \boldsymbol{\phi}_t^\star\| \leq \alpha\epsilon\right\} \geq 1 - C_t e^{-K\eta}. \tag{3.25}$$

Since $L_t^\star(\cdot)$ is also H-smooth, combing (3.24) and (3.25) proves Theorem 3.9. □

Theorem 3.9 sheds light on how the task similarity and local sample size impact the learning performance at the target edge node t. In particular, the performance gap between the optimal model and the model after fast adaptation is upper bounded by the surrogate difference, denoted by $\|\boldsymbol{\theta}_t^\star - \boldsymbol{\theta}_c^\star\|$, which serves as a guidance for the platform to determine how similar the source edge nodes in the federated meta-learning should be with the target node in order to achieve given learning performance via fast adaptation and hence edge intelligence at the target edge node.

3.6   ROBUST FEDERATED META-LEARNING (FedML)

It has been shown in Yin et al. [2018] and Edmunds et al. [2017] that meta-learning algorithms (e.g., MAML) are vulnerable to adversarial attacks, leading to possible significant performance degradation of the locally fast adapted model at the target when facing perturbed data inputs. Thus, motivated, we next devise a robust FedML algorithm and study the trade-off between robustness and accuracy therein (cf. Tsipras et al. [2018]).

3.6.1   ROBUST FEDERATED META-LEARNING

To combat against the possible vulnerability of meta-learning algorithms, we propose obtaining a model initialization from which the model updated with local training data at the target not only is robust against data distributions that are distance π away from the target data distribution P_t, but also guarantees good performance when fed with the clean target data. Based on recent advances in distributionally robust optimization (DRO), this can be achieved by solving the following problem:

$$\min_{\boldsymbol{\theta}} \left\{ L_t(\boldsymbol{\phi}_t) + \max_{P:D(P,P_t)\leq\pi} \mathbb{E}_P[l(\boldsymbol{\phi}_t, (\mathbf{x}, \mathbf{y}))] \right\}, \tag{3.26}$$

where $D(P, P_t)$ is a distance metric on the space of probability distributions.

To solve (3.26) with FedML, based on the general machine learning principle that train and test conditions must match, we can reformulate the FedML objective (3.4) as:

$$
\begin{aligned}
\min_{\boldsymbol{\theta}} \quad & \sum_{i \in \mathcal{S}} \omega_i F_i(\boldsymbol{\phi}_i), \\
\text{s.t.} \quad & F_i(\boldsymbol{\phi}_i) = L(\boldsymbol{\phi}_i, D_i^{test}) + \max_{P} \mathbb{E}_P[l(\boldsymbol{\phi}_i, (\mathbf{x}_i, \mathbf{y}_i))], \\
& \boldsymbol{\phi}_i = \boldsymbol{\theta} - \alpha \nabla_{\boldsymbol{\theta}} L(\boldsymbol{\theta}, D_i^{train}), \\
& D(P, P_i) \leq \pi,
\end{aligned}
\tag{3.27}
$$

where a distributionally robust objective function similar to the target node is set for every source edge node $i \in \mathcal{S}$.

3.6.2 WASSERSTEIN DISTANCE-BASED ROBUST FEDERATED META-LEARNING

The choice of the distributional distance metric clearly affects both the richness of the uncertainty set and the tractability of problem (3.27). To enable distance measure between distributions with different support, we use Wasserstein distance as the distance metric on the space of probability distributions. More specifically, let the transportation cost $c : (\mathcal{X} \times \mathcal{Y}) \times (\mathcal{X} \times \mathcal{Y}) \to \mathbb{R}_+$ be lower-semicontinuous and satisfy $c((\mathbf{x}, \mathbf{y}), (\mathbf{x}, \mathbf{y})) = 0$, which quantifies the cost of transporting unit mass from $(\mathbf{x}, \mathbf{y})$ to $(\mathbf{x}', \mathbf{y}')$. For any two probability measure P and Q supported on $\mathcal{X} \times \mathcal{Y}$, let $\Pi(P, Q)$ denote the set of all couplings (transport plans) between P and Q, meaning measures W with $W(A, \mathcal{X} \times \mathcal{Y}) = P(A)$ and $W(\mathcal{X} \times \mathcal{Y}, A) = Q(A)$. The Wasserstein distance is then defined as

$$
D_w(P, Q) \triangleq \inf_{W \in \Pi(P,Q)} \mathbb{E}_W[c((\mathbf{x}, \mathbf{y}), (\mathbf{x}', \mathbf{y}'))],
\tag{3.28}
$$

which represents the optimal transport cost, i.e., the lowest expected transport cost, that is achievable among all couplings between P and Q.

Since the Wasserstein distance based optimization problem is computationally demanding for complex models, based on Sinha et al. [2017], we consider the following Lagrangian relaxation of the inner maximization problem of (3.27) with penalty parameter $\lambda \geq 0$:

$$
\max_{P} \{\mathbb{E}_P[l(\boldsymbol{\phi}_i, (\mathbf{x}_i, \mathbf{y}_i))] - \lambda D_w(P, P_i)\},
\tag{3.29}
$$

where λ is inversely proportional to π. The duality result below in Blanchet and Murthy [2019] based on Kantorovich's duality, a widely used approach to solve the Wasserstein distance-based DRO problem in optimal transport, provides us an efficient way to solve (3.29) through a robust surrogate loss.

Lemma 3.10 Let $l : \mathbb{R}^d \times (\mathcal{X} \times \mathcal{Y}) \to \mathbb{R}$ and $c : (\mathcal{X} \times \mathcal{Y}) \times (\mathcal{X} \times \mathcal{Y}) \to \mathbb{R}_+$ be continuous. Define the robust surrogate loss as $l_\lambda(\boldsymbol{\theta}, (\mathbf{x}_0, \mathbf{y}_0)) \triangleq sup_{\mathbf{x} \in \mathcal{X}} \{l(\boldsymbol{\theta}, (\mathbf{x}, \mathbf{y}_0)) -$

$\lambda c((\mathbf{x}, \mathbf{y}_0), (\mathbf{x}_0, \mathbf{y}_0))\}$. For any distribution Q and $\lambda \geq 0$, we have

$$\max_{P}\{\mathbb{E}_P[l(\boldsymbol{\theta}, (\mathbf{x}, \mathbf{y}))] - \lambda D_w(P, Q)\} = \mathbb{E}_Q[l_\lambda(\boldsymbol{\theta}, (\mathbf{x}, \mathbf{y}))].$$

Lemma 3.10 divulges a worst-case joint probability measure $W^\star$ corresponding to a transport plan that transports mass from $\mathbf{x}$ to the optimizer of the local optimization problem $\sup_{\mathbf{x} \in \mathcal{X}}\{l(\boldsymbol{\theta}, (\mathbf{x}, \mathbf{y}_0)) - \lambda c((\mathbf{x}, \mathbf{y}_0), (\mathbf{x}_0, \mathbf{y}_0))\}$. Hence, we can replace (3.29) with the expected robust surrogate loss $\mathbb{E}_{P_i}[l_\lambda(\boldsymbol{\phi}_i, (\mathbf{x}_i, \mathbf{y}_i))]$. Moreover, we typically replace P_i by the empirical distribution $\hat{P}_i$ because P_i is unknown.

In what follows, we focus on the following relaxed robust problem:

$$\min_{\boldsymbol{\theta}} \quad \sum_{i \in \mathcal{S}} \omega_i\{L(\boldsymbol{\phi}_i, D_i^{test}) + \mathbb{E}_{\hat{P}_i}[l_\lambda(\boldsymbol{\phi}_i, (\mathbf{x}_i, \mathbf{y}_i))]\}, \tag{3.30}$$
$$\text{s.t.} \quad \boldsymbol{\phi}_i = \boldsymbol{\theta} - \alpha\nabla_{\boldsymbol{\theta}} L(\boldsymbol{\theta}, D_i^{train}).$$

Under suitable conditions, the robust surrogate loss $l_\lambda(\boldsymbol{\theta}, (\mathbf{x}_0, \mathbf{y}_0))$ is strongly-concave for $\lambda \geq H_\mathbf{x}$ [Sinha et al., 2017], which indicates the computational benefits of relaxing the strict robustness requirements of (3.29). Therefore,

$$\nabla_{\boldsymbol{\phi}_i} l_\lambda(\boldsymbol{\phi}_i, (\mathbf{x}_i, \mathbf{y}_i)) = \nabla_{\boldsymbol{\phi}_i} l(\boldsymbol{\phi}_i, (\mathbf{x}^\star, \mathbf{y}_i)), \tag{3.31}$$

where

$$\mathbf{x}^\star = \arg\max_{\mathbf{x} \in \mathcal{X}}\{l(\boldsymbol{\phi}_i, (\mathbf{x}, \mathbf{y}_i)) - \lambda c((\mathbf{x}, \mathbf{y}_i), (\mathbf{x}_i, \mathbf{y}_i))\}. \tag{3.32}$$

Here, $\mathbf{x}^\star$ can be regarded as an adversarial perturbation of $\mathbf{x}_i$ under current model $\boldsymbol{\phi}_i$ and efficiently approximated by gradient-ascent methods, revealing that problem (3.30) can be efficiently solved by gradient-based methods.

3.6.3   ROBUST META-TRAINING ACROSS EDGE NODES

To solve problem (3.30), similar to Volpi et al. [2018], we use an adversarial data generation process, i.e., approximately solving (3.32) with gradient ascent, to the FedML algorithm. More specifically, for every $N_0 T_0$ iterations, each edge node i constructs adversarial data samples using T_a steps gradient ascent and adds them to its own adversarial dataset D_i^{adv}. Note that this sample construction procedure can only be repeated up to R times considering the local computational resources constraints. For $t \neq nT_0$ (no global aggregation), each node i first updates $\boldsymbol{\theta}_i^t$ using the training dataset:

$$\boldsymbol{\phi}_i^t = \boldsymbol{\theta}_i^t - \alpha\nabla_{\boldsymbol{\theta}} L(\boldsymbol{\theta}_i^t, D_i^{train}), \tag{3.33}$$

then locally updates $\boldsymbol{\theta}_i^t$ again using both the testing dataset and the constructed adversarial dataset:

$$\boldsymbol{\theta}_i^{t+1} = \boldsymbol{\theta}_i^t - \beta\nabla_{\boldsymbol{\theta}}\{L(\boldsymbol{\phi}_i^t, D_i^{test}) + L(\boldsymbol{\phi}_i^t, D_i^{adv})\}. \tag{3.34}$$

Algorithm 3.2 Robust FedML

Inputs: K, T, T_0, T_a, N_0, R, α, β, ν, ω_i for $i \in \mathcal{S}$

Outputs: Final model parameter θ

1: Platform randomly initializes θ^0 and sends it to all nodes in $\mathcal{S}$; For each node i, $D_i^{adv} \leftarrow \emptyset$ and $r \leftarrow 0$;

2: **for** $t = 1, 2, \ldots, T$ **do**

3:     **for** each node $i \in \mathcal{S}$ **do**

4:         $D_i^{comb} \leftarrow D_i^{test} \cup D_i^{adv}$;

5:         Compute the updated parameter with one-step gradient descent using D_i^{train}: $\phi_i^t = \theta_i^t - \alpha \nabla_\theta L(\theta_i^t, D_i^{train})$;

6:         Obtain θ_i^{t+1} based on (3.34); **//local update**

7:         **if** $t \bmod T_0 = 0$ **then**

8:             Send θ_i^{t+1} to the platform;

9:             Receive θ^{t+1} from the platform where θ^{t+1} is obtained based on (3.6);

10:           Set $\theta_i^t \leftarrow \theta^{t+1}$; **//global aggregation**

11:         **else**

12:            Set $\theta_i^t \leftarrow \theta_i^{t+1}$;

13:         **if** $t \bmod N_0 T_0 = 0$ and $r < R$ **then //adversarial data generation**

14: Uniformly sample $(x_i^j, y_i^j)_{j=1,\ldots,|D_i^{test}|}$ from D_i^{comb}

15:         **for** $j = 1, \ldots, |D_i^{test}|$ **do**

16:            $x_i^{jr} \leftarrow x_i^j$;

17:            **for** $t = 1, \ldots, T_a$ **do**

18:              $x_i^{jr} \leftarrow x_i^{jr} + \nu \nabla_x \{l(\phi_i^t, (x_i^{jr}, y_i^j)) - \lambda c((x_i^{jr}, y_i^j), (x_i^j, y_i^j))\}$;

19:            Append (x_i^{jr}, y_i^j) to D_i^{adv};

20:         $r \leftarrow r + 1$;

21: **return** θ.

When $t = nT_0$, each node transmits the updated θ_i^{t+1} for the global aggregation (3.6). The details are summarized in Algorithm 3.2.

3.6.4 CONVERGENCE ANALYSIS

Similar with the analysis procedure for FedML, for clarity we rewrite problem (3.30) as the following:

$$\min_{\theta} \quad \tilde{G}(\theta), \tag{3.35}$$

where $\tilde{G}(\theta) = \sum_{i \in \mathcal{S}} \omega_i \tilde{G}_i(\theta)$ and $\tilde{G}_i(\theta) = L(\phi_i(\theta), D_i^{test}) + \mathbb{E}_{P_i}[l_\lambda(\phi_i(\theta), (x_i, y_i))]$.

Assumption 3.11   The function c is continuous. And for every $(\mathbf{x}_0, \mathbf{y}_0) \in \mathcal{X} \times \mathcal{Y}$, $c((\mathbf{x}, \mathbf{y}_0), (\mathbf{x}_0, \mathbf{y}_0))$ is 1-strongly convex with respect to $\mathbf{x}$.

Assumption 3.12   The loss function $l : \mathbb{R}^d \times (\mathcal{X} \times \mathcal{Y}) \to \mathbb{R}$ is μ-strongly convex with respect to θ.

Assumption 3.13   The loss function $l : \mathbb{R}^d \times (\mathcal{X} \times \mathcal{Y}) \to \mathbb{R}$ is smooth with respect to both θ and $\mathbf{x}$, i.e.,

$$\|\nabla_\theta l(\theta, (\mathbf{x}, \mathbf{y})) - \nabla_\theta l(\theta', (\mathbf{x}, \mathbf{y}))\| \leq H \|\theta - \theta'\|,$$
$$\|\nabla_\theta l(\theta, (\mathbf{x}, \mathbf{y})) - \nabla_\theta l(\theta, (\mathbf{x}', \mathbf{y}))\| \leq H_{\theta \mathbf{x}} \|\mathbf{x} - \mathbf{x}'\|,$$
$$\|\nabla_\mathbf{x} l(\theta, (\mathbf{x}, \mathbf{y})) - \nabla_\mathbf{x} l(\theta, (\mathbf{x}', \mathbf{y}))\| \leq H_{\mathbf{xx}} \|\mathbf{x} - \mathbf{x}'\|,$$
$$\|\nabla_\mathbf{x} l(\theta, (\mathbf{x}, \mathbf{y})) - \nabla_\mathbf{x} l(\theta', (\mathbf{x}, \mathbf{y}))\| \leq H_{\mathbf{x}\theta} \|\theta - \theta'\|,$$

and there exists a constant B such that $\|\nabla_\theta l(\theta, (\mathbf{x}, \mathbf{y}))\| \leq B$ for all $\theta \in \mathbb{R}^d$ and $(\mathbf{x}, \mathbf{y}) \in \mathcal{X} \times \mathcal{Y}$.

Note that Assumptions 3.12–3.13 can be stronger replacements of Assumptions 3.1–3.2, respectively. We characterize the robust FedML objective $\tilde{G}(\theta)$ below.

Theorem 3.14   *Suppose Assumptions 3.3 and 3.11–3.13 hold. When $\alpha \leq \min\{\frac{\mu}{2\mu H + \rho B}, \frac{1}{\mu}\}$ and $\lambda \geq H_{\mathbf{xx}} + \frac{H_{\theta \mathbf{x}} H_{\mathbf{x}\theta}}{\mu}$, problem (3.35) has a unique minimizer.*

Proof. Let $\tilde{l}(\theta, \mathbf{x}) \triangleq l(\theta, (\mathbf{x}, \mathbf{y}_0)) - \lambda c((\mathbf{x}, \mathbf{y}_0), (\mathbf{x}_0, \mathbf{y}_0))$, and then the robust surrogate loss $l_\lambda(\theta, (\mathbf{x}_0, \mathbf{y}_0)) = \sup_{\mathbf{x} \in \mathcal{X}} \tilde{l}(\theta, \mathbf{x})$. Based on Assumptions 5 and 7, it can be easily shown that $\tilde{l}(\theta, \mathbf{x})$ is $(\lambda - H_{\mathbf{xx}})$-strongly concave with respect to $\mathbf{x}$. Let $\mathbf{x}^\star(\theta) = \arg\max_{\mathbf{x} \in \mathcal{X}} \tilde{l}(\theta, \mathbf{x})$. From Lemma 1 in Sinha et al. [2017], we conclude that l_λ is differentiable and $\nabla_\theta l_\lambda(\theta, (\mathbf{x}_0, \mathbf{y}_0)) = \nabla_\theta \tilde{l}(\theta, \mathbf{x}^\star(\theta))$. Moreover,

$$\|\mathbf{x}^\star(\theta) - \mathbf{x}^\star(\theta')\| \leq \frac{H_{\mathbf{x}\theta}}{\lambda - H_{\mathbf{xx}}} \|\theta - \theta'\|, \tag{3.36}$$

and

$$\|\nabla_\theta l_\lambda(\theta) - \nabla_\theta l_\lambda(\theta')\| \leq \left(H + \frac{H_{\theta \mathbf{x}} H_{\mathbf{x}\theta}}{\lambda - H_{\mathbf{xx}}} \right) \|\theta - \theta'\| \triangleq H_\lambda \|\theta - \theta'\|.$$

Further, we have that

$$\|\nabla_{\boldsymbol{\theta}} l_\lambda(\boldsymbol{\theta}) - \nabla_{\boldsymbol{\theta}} l_\lambda(\boldsymbol{\theta}')\|$$
$$=\|\nabla_{\boldsymbol{\theta}} \tilde{l}(\boldsymbol{\theta}, \mathbf{x}^\star(\boldsymbol{\theta})) - \nabla_{\boldsymbol{\theta}} \tilde{l}(\boldsymbol{\theta}', \mathbf{x}^\star(\boldsymbol{\theta}'))\|$$
$$=\|\nabla_{\boldsymbol{\theta}} \tilde{l}(\boldsymbol{\theta}, \mathbf{x}^\star(\boldsymbol{\theta})) - \nabla_{\boldsymbol{\theta}} \tilde{l}(\boldsymbol{\theta}, \mathbf{x}^\star(\boldsymbol{\theta}')) + \nabla_{\boldsymbol{\theta}} \tilde{l}(\boldsymbol{\theta}, \mathbf{x}^\star(\boldsymbol{\theta}')) - \nabla_{\boldsymbol{\theta}} \tilde{l}(\boldsymbol{\theta}', \mathbf{x}^\star(\boldsymbol{\theta}'))\|$$
$$\geq\|\nabla_{\boldsymbol{\theta}} \tilde{l}(\boldsymbol{\theta}, \mathbf{x}^\star(\boldsymbol{\theta}')) - \nabla_{\boldsymbol{\theta}} \tilde{l}(\boldsymbol{\theta}', \mathbf{x}^\star(\boldsymbol{\theta}'))\| - \|\nabla_{\boldsymbol{\theta}} \tilde{l}(\boldsymbol{\theta}, \mathbf{x}^\star(\boldsymbol{\theta})) - \nabla_{\boldsymbol{\theta}} \tilde{l}(\boldsymbol{\theta}, \mathbf{x}^\star(\boldsymbol{\theta}'))\|$$
$$\geq\mu\|\boldsymbol{\theta} - \boldsymbol{\theta}'\| - H_{\boldsymbol{\theta}\mathbf{x}}\|\mathbf{x}^\star(\boldsymbol{\theta}) - \mathbf{x}^\star(\boldsymbol{\theta}')\|$$
$$\geq \left(\mu - \frac{H_{\boldsymbol{\theta}\mathbf{x}} H_{\mathbf{x}\boldsymbol{\theta}}}{\lambda - H_{\mathbf{x}\mathbf{x}}}\right)\|\boldsymbol{\theta} - \boldsymbol{\theta}'\| \triangleq \mu_\lambda\|\boldsymbol{\theta} - \boldsymbol{\theta}'\|.$$

So the strongly convexity and smoothness of the robust surrogate loss l_λ still hold if λ is large enough. Based on the triangle inequality, $\mathbb{E}_{P_i}[l_\lambda(\boldsymbol{\theta}, (\mathbf{x}_i, \mathbf{y}_i))]$ is μ_λ-strongly convex and H_λ-smooth both with respect to $\boldsymbol{\theta}$ for all $i \in \mathcal{S}$.

Denote $\tilde{H}_i(\boldsymbol{\theta}) = \mathbb{E}_{P_i}[l_\lambda(\boldsymbol{\theta}, (\mathbf{x}_i, \mathbf{y}_i))]$. Then we have

$$\|\nabla_{\boldsymbol{\theta}} \mathbb{E}_{P_i}[l_\lambda(\boldsymbol{\phi}_i(\boldsymbol{\theta}), (\mathbf{x}_i, \mathbf{y}_i))] - \nabla_{\boldsymbol{\theta}} \mathbb{E}_{P_i}[l_\lambda(\boldsymbol{\phi}_i'(\boldsymbol{\theta}'), (\mathbf{x}_i, \mathbf{y}_i))]\|$$
$$=\|\nabla \tilde{H}_i(\boldsymbol{\phi}_i)[I - \alpha\nabla^2 L_i(\boldsymbol{\theta})] - \nabla \tilde{H}_i(\boldsymbol{\phi}_i')[I - \alpha\nabla^2 L_i(\boldsymbol{\theta}')]\|$$
$$=\|\nabla \tilde{H}_i(\boldsymbol{\phi}_i)[I - \alpha\nabla^2 L_i(\boldsymbol{\theta})] - \nabla \tilde{H}_i(\boldsymbol{\phi}_i')[I - \alpha\nabla^2 L_i(\boldsymbol{\theta}')] + \alpha\nabla \tilde{H}_i(\boldsymbol{\phi}_i')\nabla^2 L_i(\boldsymbol{\theta}) - \alpha\nabla \tilde{H}_i(\boldsymbol{\phi}_i')\nabla^2 L_i(\boldsymbol{\theta})\|$$
$$=\|[\nabla H_i(\boldsymbol{\phi}_i) - \nabla \tilde{H}_i(\boldsymbol{\phi}_i')][I - \alpha\nabla^2 L_i(\boldsymbol{\theta})] - \alpha\nabla \tilde{H}_i(\boldsymbol{\phi}_i')[\nabla^2 L_i(\boldsymbol{\theta}) - \nabla^2 L_i(\boldsymbol{\theta}')]\|. \tag{3.37}$$

Since $\nabla\boldsymbol{\phi}_i = I - \alpha\nabla^2 L_i(\boldsymbol{\theta})$, it follows from Assumptions 3.12 and 3.13 that $1 - \alpha H \leq \nabla\boldsymbol{\phi}_i \leq 1 - \alpha\mu$, which indicates

$$(1 - \alpha H)\|\boldsymbol{\theta} - \boldsymbol{\theta}'\| \leq \|\boldsymbol{\phi}_i - \boldsymbol{\phi}_i'\| \leq (1 - \alpha\mu)\|\boldsymbol{\theta} - \boldsymbol{\theta}'\|.$$

Continuing with (3.37), it is clear that

$$\|\nabla_{\boldsymbol{\theta}} \mathbb{E}_{P_i}[l_\lambda(\boldsymbol{\phi}_i(\boldsymbol{\theta}), (\mathbf{x}_i, \mathbf{y}_i))] - \nabla_{\boldsymbol{\theta}} \mathbb{E}_{P_i}[l_\lambda(\boldsymbol{\phi}_i'(\boldsymbol{\theta}'), (\mathbf{x}_i, \mathbf{y}_i))]\|$$
$$\geq(1 - \alpha H)\|\nabla \tilde{H}_i(\boldsymbol{\phi}_i) - \nabla \tilde{H}_i(\boldsymbol{\phi}_i')\| - \alpha B\|\nabla^2 L_i(\boldsymbol{\theta}) - \nabla^2 L_i(\boldsymbol{\theta}')\|$$
$$\geq\mu_\lambda(1 - \alpha H)\|\boldsymbol{\phi}_i - \boldsymbol{\phi}_i'\| - \alpha B\rho\|\boldsymbol{\theta} - \boldsymbol{\theta}'\|$$
$$\geq[\mu_\lambda(1 - \alpha H)^2 - \alpha B\rho]\|\boldsymbol{\theta} - \boldsymbol{\theta}'\|,$$

and

$$\|\nabla_{\boldsymbol{\theta}} \mathbb{E}_{P_i}[l_\lambda(\boldsymbol{\phi}_i(\boldsymbol{\theta}), (\mathbf{x}_i, \mathbf{y}_i))] - \nabla_{\boldsymbol{\theta}} \mathbb{E}_{P_i}[l_\lambda(\boldsymbol{\phi}_i'(\boldsymbol{\theta}'), (\mathbf{x}_i, \mathbf{y}_i))]\|$$
$$\leq(1 - \alpha\mu)\|\nabla \tilde{H}_i(\boldsymbol{\phi}_i) - \nabla \tilde{H}_i(\boldsymbol{\phi}_i')\| + \alpha B\|\nabla^2 L_i(\boldsymbol{\theta}) - \nabla^2 L_i(\boldsymbol{\theta}')\|$$
$$\leq H_\lambda(1 - \alpha\mu)\|\boldsymbol{\phi}_i - \boldsymbol{\phi}_i'\| + \alpha B\rho\|\boldsymbol{\theta} - \boldsymbol{\theta}'\|$$
$$\leq[H_\lambda(1 - \alpha\mu)^2 + \alpha B\rho]\|\boldsymbol{\theta} - \boldsymbol{\theta}'\|.$$

From the definition of $\tilde{G}_i(\boldsymbol{\theta})$, we can have that $\tilde{G}_i(\boldsymbol{\theta})$ is μ_R-strongly convex and H_R-smooth, where $\mu_R = \mu + \mu_\lambda(1 - \alpha H)^2 - \alpha B\rho$ and $H_R = H + H_\lambda(1 - \alpha\mu)^2 + \alpha B\rho$. Based on the triangle inequality, Theorem 3.14 can be proved. $\qquad\square$

Theorem 3.14 implies that when the learning rate α is sufficiently small and the Lagrangian penalty parameter λ is large enough, the relaxed robust meta-learning objective function $\tilde{G}(\theta)$ is strongly convex and hence has a unique solution. Further, as outlined in Algorithm 3.2, through pre-training each edge node to learn to protect against adversarial perturbations on the testing dataset while securing the model accuracy on clean data with Algorithm 3.2, the learned model via meta-training automatically gains the ability to prevent future adversarial attacks without significantly sacrificing the learning accuracy with quick adaptation at the target edge node.

3.7 EXPERIMENTS

In this section, we first introduce the experimental setup, and then evaluate the performance of FedML and Robust FedML. In particular, we investigate the impact of node similarity and number of local update step T_0 on the convergence of FedML, and compare the fast adaptation performance of FedML with the model learned from federated learning (Fedavg) [McMahan et al., 2016].

3.7.1 EXPERIMENTAL SETTING

Synthetic data. To evaluate the impact of node similarity on the performance of convergence and fast adaptation, we follow a similar setup in Sahu et al. [2018] to generate synthetic data. Specifically, the synthetic sample $(\mathbf{x}_i^j, \mathbf{y}_i^j)$ for each node i is generated from the model $\mathbf{y} = \arg\max(\text{softmax}(\mathbf{Wx} + \mathbf{b}))$ where $\mathbf{x} \in \mathbb{R}^{60}$, $\mathbf{W} \in \mathbb{R}^{10 \times 60}$ and $\mathbf{b} \in \mathbb{R}^{10}$. Moreover, $\mathbf{W}_i \sim \mathcal{N}(\mathbf{u}_i, 1)$, $\mathbf{b}_i \sim \mathcal{N}(\mathbf{u}_i, 1)$, $\mathbf{u}_i \sim \mathcal{N}(0, \tilde{\alpha})$; $\mathbf{x}_i^j \sim \mathcal{N}(\mathbf{v}_i, \Sigma)$ where the covariance matrix Σ is diagonal with $\Sigma_{k,k} = k^{-1.2}$ and $\mathbf{v}_i \sim \mathcal{N}(B_i, 1)$, $B_i \sim N(0, \tilde{\beta})$. Intuitively, $\tilde{\alpha}$ and $\tilde{\beta}$ control the local model similarity across all nodes, which can be changed to generate heterogeneous local datasets named Synthetic($\tilde{\alpha}, \tilde{\beta}$). For all synthetic datasets, we consider 50 nodes in total and the number of samples on each node follows a power law. The objective is to learn the model parameters $\mathbf{W}$ and $\mathbf{b}$ with the cross-entropy loss function.

Real data. We also explore two real datasets, MNIST [LeCun et al., 1998] and Sentiment140 (Sent140) [Go et al., 2009], used for text sentiment analysis on tweets. For MNIST, we sample part of data and distribute the data among 100 nodes such that every node has samples of only two digits and the number of samples per device follows a power law. We study a convex classification problem with MNIST using multinomial logistic regression. Next, we consider a more complicated classification problem on Sent140 by taking each twitter account as a node, where the model takes a sequence of 25 characters as input, embeds each of the character into a 300-dimensional space by looking up the pretrained 300D GloVe embedding [Pennington et al., 2014], and outputs one character per training sample through a network with 3 hidden layers with sizes 256, 128, 64, each including batch normalization and ReLU nonlinearities, followed

Table 3.1: Statistics of datasets

Dataset	Nodes	Sample Per Node	
		Mean	Stdev
Synthetic	50	17	5
MNIST	100	34	5
Sent140	706	42	35

by a linear layer and softmax. The loss function is the cross-entropy error between the predicted and true class for all models.

Implementation. For each node, we divide the local dataset as a training set and a testing set. We select 80% nodes as the source nodes and evaluate the fast adaptation performance on the rest. When training with FedML, we vary the size of the training set, i.e., K, for the one-step gradient update, whereas the entire dataset is used for training in Fedavg. During testing, the trained model is first updated with the training set of testing nodes, and then evaluated on their testing sets. For FedML, we set both the learning rate α and meta learning rate β as 0.01 for synthetic data and MNIST, while $\alpha = 0.01$ and $\beta = 0.3$ for Sent140. Fedavg has the same learning rate with β.

3.7.2 EVALUATION OF FEDERATED META-LEARNING

Convergence behavior. We evaluate the convergence error for (a) three different synthetic datasets with $T_0 = 10$ and (b) the same dataset but different T_0 with $T = 500$. As illustrated in Figure 3.2, the experimental results corroborate Theorem 3.7 that the convergence error of FedML decreases with the node similarity but increases with T_0 given a fixed algorithm duration T. Moreover, the result in Sent140 (Figure 3.3) shows that FedML also achieves good convergence performance in practical non-convex settings.

Fast adaptation performance. As shown in Figure 3.4a, FedML achieves the best adaptation performance on Synthetic(0,0) where the nodes are the most similar. We also compare the fast adaptation performance between FedML and Fedavg on three different datasets, Synthetic(0.5,0.5), MNIST, and Sent140. As shown in Figures 3.4b–d, the model learned from FedML can achieve significantly better adaptation performance at the target nodes compared with that in Fedavg, and this performance gap increases when the target node has small local datasets. It can be seen that the model learned in Fedavg turns to have overfitting issues when fine-tuned with a few data samples, whereas the meta-model in FedML improves with additional gradient steps without overfitting.

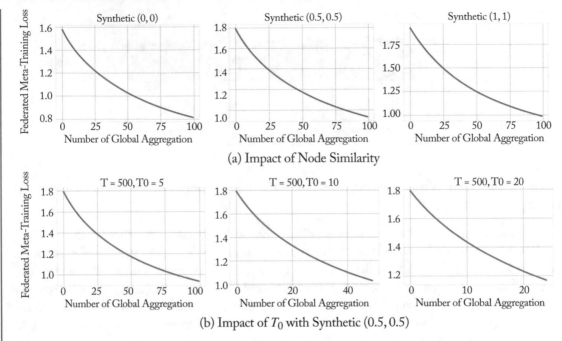

(a) Impact of Node Similarity

(b) Impact of T_0 with Synthetic $(0.5, 0.5)$

Figure 3.2: Impact of node similarity and T_0 on the convergence of FedML.

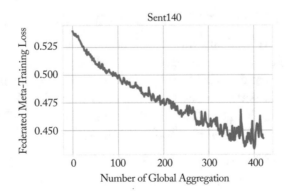

Figure 3.3: Convergence of FedML on Sent140.

3.7.3   EVALUATION OF ROBUST FEDERATED META-LEARNING

We compare the adaptation performance of FedML and Robust FedML on MNIST with $T_0 = 5$ during training. For adversarial perturbations only to feature vectors in supervised learning, we consider the transportation cost function as: $c((\mathbf{x}, \mathbf{y}), (\mathbf{x}', \mathbf{y}')) = \|\mathbf{x} - \mathbf{x}'\|_2^2 + \infty \cdot \mathbf{1}(\mathbf{y} - \mathbf{y}')$. The learning rate $v = 1$, $R = 2$, $N_0 = 7$ and $T_a = 10$ for adversarial data generation. For testing

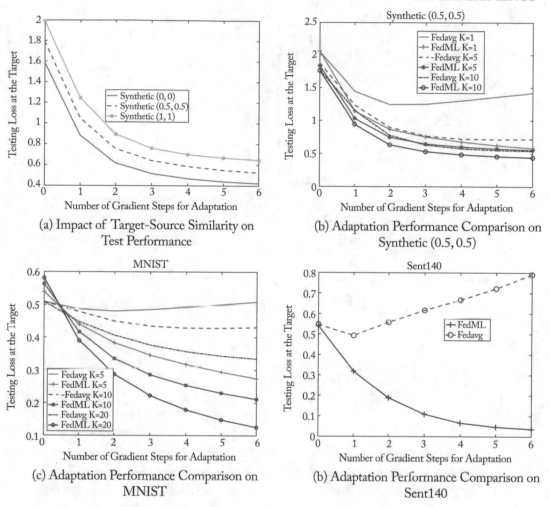

Figure 3.4: Convergence and fast adaptation performance of FedML on different datasets with $T_0 = 5$.

at the target, we first update the meta-model with clean training data, and then evaluate the adaptation performance on clean test data and adversarial data, where the adversarial data is generated by using the Fast Gradient Sign Method [Goodfellow et al., 2014b] with parameter ξ, respectively. Since the size of distributional uncertainty set controls the trade-off between robustness and accuracy, we compare the performance of Robust FedML with $\lambda = 0.1, 1, 10$, where the smaller λ is, the more robustness Robust FedML provides.

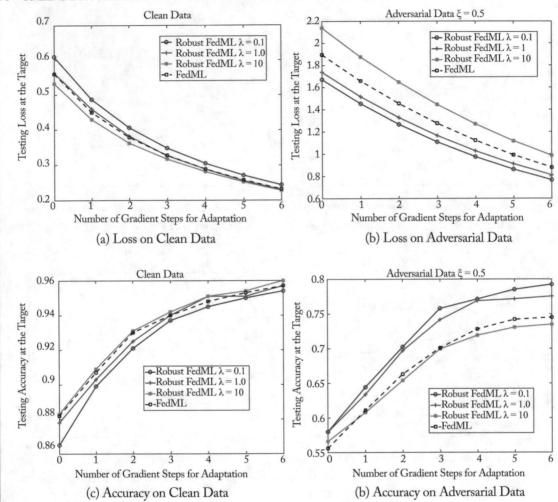

Figure 3.5: Adaptation performance of robust FedML on MNIST with $T_0 = 5$.

Robustness-Accuracy tradeoff. As shown in Figure 3.5a–d, when λ decreases, Robust FedML has slightly worse performance on clean data and the performance on adversarial data is much better. Compared with the case where $\lambda = 10$, Robust FedML with $\lambda = 0.1$ significantly improves the robustness against adversarial data without sacrificing too much on the accuracy with clean data. Moreover, Robust FedML with smaller λ is more robust than FedML. Note that the uncertainty set is too small to positively affect the robustness when $\lambda = 10$.

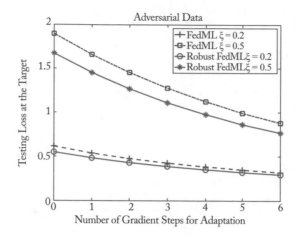

Figure 3.6: Impact of ξ.

Impact of ξ. Clearly, both FedML and Robust FedML achieve better performance when facing smaller perturbation (smaller ξ) of testing data. Figure 3.6 further indicates that the improvement of Robust FedML over FedML is higher with more perturbed data.

3.8   SUMMARY

In this chapter, we propose a platform-aided collaborative learning framework, where a model is first trained across a set of source edge nodes by a FedML approach, and then it is rapidly adapted to achieve real-time edge intelligence at the target edge node, using a few samples only. We investigate the convergence of FedML under mild conditions on node similarity, and study the adaptation performance to achieve edge intelligence at the target node. To combat against the vulnerability of meta-learning algorithms, we further propose a robust FedML algorithm based on DRO with convergence guarantee. Experimental results on various datasets corroborate the effectiveness of the proposed collaborative learning framework.

CHAPTER 4

Edge-Cloud Collaborative Learning via Distributionally Robust Optimization

4.1 INTRODUCTION

As noted before, the real-time edge intelligence can be achieved by the collaborative learning across different edge nodes. In this chapter, we turn to studying the collaborative learning between the edge and the cloud to achieve real-time edge intelligence. More specifically, we propose a distributionally robust optimization (DRO) approach which provides a natural mechanism to enable the synergy between the local data processing and the cloud knowledge transfer. In particular, we construct a distributional uncertainty set to capture the inaccuracy of local data processing, and build another distribution uncertainty model corresponding to the cloud knowledge transfer. Then, we recast the edge learning problem as a DRO problem which takes into account the two distributional uncertainty constraints. Further, since the cloud knowledge transfer hinges upon the nature of the learning problem, it may take different forms. Thus, motivated, we will investigate two interesting uncertainty models corresponding to the cloud knowledge transfer, one in terms of the distribution uncertainty set based on the cloud data distribution [Zhang et al., 2020c], and the other in terms of the prior distribution of the edge model conditioned on the cloud model [Zhang et al., 2020a,b]. We will devise collaborative learning algorithms along this line, and quantify the performance gain of the proposed DRO-enhance edge learning framework over approaches using local edge data only.

4.2 BASIC SETTING FOR COLLABORATING LEARNING TOWARD EDGE INTELLIGENCE

Consider the following learning task at an edge device:

$$\min_{\mathbf{w}_e} \mathbb{E}_{\boldsymbol{\xi} \sim P_t}[h(\mathbf{w}_e, \boldsymbol{\xi})], \tag{4.1}$$

where the edge device aims to learn the model parameter $\mathbf{w}_e$ which minimizes the cost function $h(\mathbf{w}_e, \boldsymbol{\xi})$, based on its local dataset. Denote the labeled dataset of the edge device by $\mathcal{D}_e = \{\boldsymbol{\xi}_i = (\mathbf{x}_i, y_i)\}_{i=1}^N$, with N samples of input-output pair $\boldsymbol{\xi} = (\mathbf{x}, y) \in \Xi \subset \mathbb{X} \times \mathbb{Y}$, which is assumed to follow some unknown underlying distribution P_t on $\Xi \subset \mathbb{X} \times \mathbb{Y}$. Figure 4.1 illustrates the

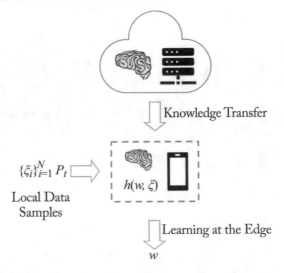

Figure 4.1: Illustration of real-time learning by the edge IoT device.

real-time learning at the edge IoT device. It is clear that the empirical distribution from the data sample set is $P_o = \frac{1}{N} \sum_{i=1}^{N} \delta_{(\mathbf{x}_i, y_i)}$, where $\delta_{(\mathbf{x}_i, y_i)}$ denotes the Dirac point measure at $\boldsymbol{\xi}_i = (\mathbf{x}_i, y_i)$.

A standard approach to solving the above edge learning problem would be to find W_e minimizing the empirical loss function based on the local data of an edge device only, i.e.,

$$\min_{\mathbf{w}_e} \mathbb{E}_{\boldsymbol{\xi} \sim P_o} [h(\mathbf{w}_e, \boldsymbol{\xi})]. \tag{4.2}$$

In general, the resulting $\mathbf{w}_e$ would exhibit poor out-of-sample performance, since the sample size of $\mathcal{D}_e$ is typically small at an individual edge device and hence P_o is distant from P_t. It is therefore of great importance to come up innovative algorithms to boost up the edge learning performance, but this intriguing problem has been open by and large.

Observe that relevant model parameters $\mathbf{w}_c$ and distributional information about the training data can be learned from the large amount of historical data in the cloud, and further the plenteous computing resources enable the cloud to obtain a more accurate model therein which is expected to be informative and hence help to train the model at the edge device. Nevertheless, the model learned in the cloud in general is not exactly the same as the edge model so it cannot be directly applied to the edge device, although it is relevant. One natural question to ask here is: "*How can we leverage the knowledge transfer from the cloud to reach a more intelligent decision making to achieve edge intelligence?*" A key challenge here is to come up with the right abstraction of the knowledge transfer that can be cohesively integrated with the local data processing, so as to improve edge learning performance. To tackle this challenge, we propose a distributionally robust optimization framework which provides a natural mechanism to enable the synergy between the local data processing and the knowledge transfer from the cloud. More specifically, observe that the data sample size at the edge device is typically small and hence P_o is distant from P_t, indicating that perfect data distributions are impossible to attain at edge nodes. It therefore makes sense to construct a distributional uncertainty set based on local data. Further, since the model learned in the cloud in general is not the same as the edge model but is informative and relevant, it is reasonable to construct another distribution uncertainty model based on cloud learning for knowledge transfer accordingly. With this insight, we recast the edge learning problem as a DRO problem which takes into account two distributional uncertainty constraints, one centered around the local empirical distribution P_o and the other based on the cloud knowledge, respectively. We note that the knowledge transfer from the cloud to the edge hinges upon the nature of the learning problem under consideration, so it may take different forms tailored for black-box deep learning, classification/regression and reinforcement learning, respectively. For convenience, let $P_c(\mathbf{x}, y)$ denote the distribution of the labeled data in the cloud (e.g., $P_c(\mathbf{x}, y)$ can be obtained from the generator in a GAN pre-trained in the cloud) and $P(\mathbf{w}_e|\mathbf{w}_c)$ denote the prior distribution of $\mathbf{w}_e$ conditioned on $\mathbf{w}_c$. We will investigate two interesting uncertainty models corresponding to the cloud knowledge transfer, one in terms of the distribution uncertainty set based on $P_c(\mathbf{x}, y)$, and the other in terms of the conditional prior distribution $P(\mathbf{w}_e|\mathbf{w}_c)$, as illustrated in Figure 4.2. In a nutshell, it is anticipated that the edge device is able to improve the learning performance, by leveraging the cloud-edge synergy under the DRO framework.

4.3   COLLABORATIVE LEARNING BASED ON EDGE-CLOUD SYNERGY OF DISTRIBUTION UNCERTAINTY SETS

In this section, we present a DRO-based edge intelligence framework for achieving real-time decision making, in which we formulate the edge learning as a DRO problem subject to two distributional uncertainty sets, one centered around the local empirical distribution and the other around the cloud reference distribution, respectively.

4.3.1   CLOUD KNOWLEDGE TRANSFER BASED ON EDGE-CLOUD MODEL RELATION

With the abundant computing resources and massive data amount at the cloud, it is reasonable to assume that the cloud has the capability to learn with enough accuracy a reference distribution P_c of the random vector $\boldsymbol{\xi}_c$ and make an optimal decision $\mathbf{w}_c$ corresponding to the cloud dataset $\mathcal{D}_c = \{\boldsymbol{\xi}_1^c, \boldsymbol{\xi}_2^c, \dots, \boldsymbol{\xi}_{N_c}^c\}$, where $N_c >> N$.

We further assume that the reference distribution P_c can be parameterized by $\mathbf{w}_c \in \mathbb{R}^k$. Following Karbalayghareh et al. [2018], for better understanding the "transferability" between the cloud and the network edge, we assume that P_c and P_t are related through a joint prior

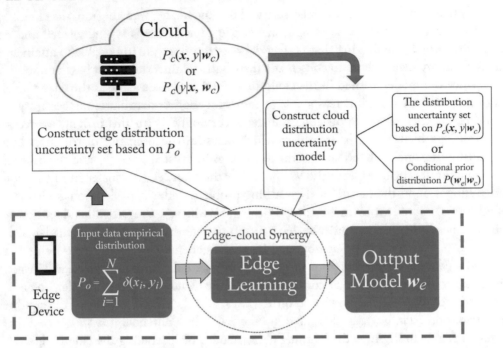

Figure 4.2: Illustration of constructing distribution uncertainty models toward edge-cloud synergy.

distribution of the model parameters $\mathbf{w}_c$ and $\mathbf{w}_t$, i.e., $(\mathbf{w}_c, \mathbf{w}_t) \sim P_\mathbf{w}$. In other words, this joint distribution of the model parameters represents the dependency between the cloud and the edge domains, which indicates that it is meaningful and critical for the edge nodes to incorporate the knowledge transferred from the cloud.

4.3.2   DRO FORMULATION WITH TWO DISTRIBUTION DISTANCE-BASED UNCERTAINTY SETS

Recall that due to the limited data at the edge device, P_o is different from P_t, and hence it makes sense to construct a distributional uncertainty set based on local data. With this insight, we construct the distribution distance based uncertainty set as the collection of distributions of $(\mathbf{x}, y)$ that are contained in a distributional "ball" centered around the local empirical distribution P_o, given below:

$$D(P, P_c) \leq \eta_1. \tag{4.3}$$

The knowledge transferred from the cloud to the edge device is in the form of the reference distribution P_c and its associated uncertainty set, given as follows:

$$D(P, P_c) \leq \eta_2, \tag{4.4}$$

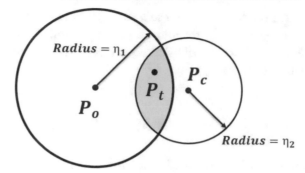

Figure 4.3: Illustration of two distribution distance-based uncertainty sets centered around the empirical distribution P_o and the cloud reference distribution P_c, respectively.

where the distribution distance based uncertainty set is constructed in a way that the distributions of $P_c(\mathbf{x}, y)$ are contained in a distributional "ball" centered around the cloud reference distribution P_c with radius η_2.

Combining Problem (4.2) with the two uncertainty sets in (4.3) and (4.4), we recast the edge learning problem as a distributionally robust optimization problem, outlined as follows:

$$\min_{\mathbf{w}_e} \max_{P} \mathbb{E}_{\xi \sim P}[h(\mathbf{w}_e, \xi)] \tag{4.5}$$

$$\text{s.t.} \ \ D(P, P_o) \leq \eta_1; \quad D(P, P_c) \leq \eta_2.$$

As shown in Figure 4.3, with careful selections of η_1 and η_2, the uncertainties at the edge can be reduced such that a smaller uncertainty set could be obtained for better estimation of the underlying distribution P_t. The performance of the proposed edge intelligence framework highly depends on the values of η_1 and η_2. Intuitively, the parameters η_1 and η_2 provide natural means for quantifying the confidence in the edge local data samples and the cloud reference distribution P_c, respectively.

Following the literature on DRO, *chance constrainted programs* can be applied to determine η_1 and η_2 such that the two distribution uncertainty sets contain P_t with certain confidence level $1 - \beta_1$ and $1 - \beta_2$ for some prescribed $\beta_1 \in (0, 1)$ and $\beta_2 \in (0, 1)$, respectively. Specifically, we have that

$$\mathbb{P}\left(D(P, P_o) \leq \eta_1\right) \geq 1 - \beta_1, \tag{4.6}$$

$$\mathbb{P}\left(D(P, P_c) \leq \eta_2\right) \geq 1 - \beta_2. \tag{4.7}$$

To evaluate η_1 and η_2 in terms of the corresponding confidence levels, it suffices to characterize the distributions of two distances $D(P, P_o)$ and $D(P, P_c)$. Specifically, by utilizing the statistical properties of the sampled data at the edge device, the distribution of $D(P, P_o)$ can be derived as a function of the data sample number N, and consequently η_1 is a function of both N

and β_1 (we elaborate further on this in the next subsection). For the selection of η_2, the marginal distribution of $\mathbf{w}_t$ can be determined by the joint prior distribution $P_{\mathbf{w}}$ of the model parameters $\mathbf{w}_c$ and $\mathbf{w}_t$ when the cloud reference distribution P_c is given. The distribution of $D(P, P_c)$ can be represented as a function of $\mathbf{w}_t$.

4.3.3   CONSTRUCTION OF WASSERSTEIN DISTANCE-BASED UNCERTAINTY SETS

Clearly, the construction of two distribution distance-based uncertainty sets in (4.5) is a key step in the proposed DRO-enhanced edge learning framework. A good uncertainty set should be rich enough to contain the true underlying distribution P_t, and should also be compatible with the statistical properties of the data. In this task, we use the Wasserstein metric as the distribution distance metric, which has been widely used in machine learning. We note that another popular distance metric is the KL divergence, which, nevertheless, has some limitations. In particular, if the reference distribution is a discrete empirical distribution, then this KL divergence ball would not contain continuous distributions, simply because the KL divergence ball does not include distributions with support on points where the reference distribution is not supported. This is why we choose the Wasserstein metric instead.

The Wasserstein metric [Villani, 2008] has been studied extensively in the field of optimal transport theory. Let $\mathcal{M}(\Xi)$ denote the collections of all probability measures on Ξ. The definition of the Wasserstein metric is given in the following.

Definition 4.1   Wasserstein distance.   For two probability distribution $P_1, P_2 \in \mathcal{M}(\Xi)$ in the same probability space Ξ, we define the rth Wasserstein distance between them as follows:

$$
W_r(P_1, P_2) := \left\{ \inf_{\pi \in \Pi(P_1, P_2)} \int_{\Xi \times \Xi} d^r(\xi_1, \xi_2) \, d\pi(\xi_1, \xi_2) \right\}^{1/r}, \tag{4.8}
$$

where $d(\xi_1, \xi_2)$ is an arbitrary norm function and $\Pi(P_1, P_2)$ denotes the set of all probability measures on $\Xi \times \Xi$ with marginal probabilities P_1 and P_2 on ξ_1 and ξ_2, respectively.

In the optimal transport problem, $d(\xi_1, \xi_2)$ characterizes the cost of moving a unit mass from ξ_1 to ξ_2, and the variable π represents a transportation policy that moves ξ_1 following a mass distribution P_1 to another one P_2. Intuitively, the Wasserstein distance corresponds to the cost of an optimal transportation policy.

Next, we need to choose a proper η_1 such that the underlying distribution P_t belongs to the uncertainty set based on the empirical distribution P_o with $1 - \beta_1$ confidence level. In order to compute η_1, we assume a light-tailed model for the underlying distribution P_t.

Assumption 4.2   Light-tailed distribution   Suppose the distribution P_t is light-tailed. That is, there exists a number $a > 1$ such that:

$$\mathbb{E}_{\xi \sim P_t} \left[\exp \left(\| \xi \|^a \right) \right] = \int_{\xi \in \Xi} \exp \left(\| \xi \|^a \right) P_t(\mathrm{d}\xi) < \infty. \tag{4.9}$$

The above inequality shows that the underlying distribution P_t goes to 0 faster than the exponential and has less mass in the tail. Note that light-tailed distributions are common in statistics (e.g., the normal distribution and t-distribution). The following lemma is concerned with the selection of parameter η_1.

Lemma 4.3   Measure concentration [Fournier and Guillin, 2015]   If Assumption 4.2 holds, then we have

$$\mathbb{P}\left(W\left(P_t, P_o \right) \geq \eta_1 \right) \leq \begin{cases} c_1 \exp \left(-c_2 N \eta_1^{\max\{m,2\}} \right) & \text{if } \eta_1 \leq 1, \\ c_1 \exp \left(-c_2 N \eta_1^a \right) & \text{if } \eta_1 > 1, \end{cases} \tag{4.10}$$

where $N > 1$ is the number of sampled data collected by edge, m is the dimension of the probability space Ξ, and c_1, c_2 are two positive constants only depending on a and m.

For a given confidence level $1 - \beta_1 \in (0, 1)$, according to (4.10) we can obtain

$$\eta_1 := \begin{cases} \left(\dfrac{\log \left(c_1 \beta_1^{-1} \right)}{c_2 N} \right)^{1/\max\{m,2\}} & \text{if } N \geq \dfrac{\log \left(c_1 \beta_1^{-1} \right)}{c_2}, \\[4ex] \left(\dfrac{\log \left(c_1 \beta_1^{-1} \right)}{c_2 N} \right)^{1/a} & \text{if } N < \dfrac{\log \left(c_1 \beta_1^{-1} \right)}{c_2}. \end{cases} \tag{4.11}$$

4.3.4 A DUALITY APPROACH TO THE WASSERSTEIN DISTANCE-BASED DRO FOR EDGE LEARNING

To reduce the computational complexity, we will use the celebrated Kantorovich duality representation to solve the Wasserstein Distance-based DRO problem in (4.5).

Lemma 4.4 Kantorovich duality [Kantorovich, 2006] The Wasserstein distance admits a reformulation due to Kantorovich's duality:

$$
W_1(P_1, P_2) = \sup_{\phi, \psi} \left\{ \int_{\Xi} \phi(\xi_1) P_1(d\xi_1) + \right.
$$
$$
\left. \int_{\Xi} \psi(\xi_2) P_2(d\xi_2) : \phi(\xi_1) + \psi(\xi_2) \leq d(\xi_1, \xi_2), \forall \xi_1, \xi_2 \right\}. \tag{4.12}
$$

Definition 4.5 c-transformation [Ambrosio and Gigli, 2013]. Given a function $\phi : \xi_1 \rightarrow \mathbb{R}$, then $d(\xi_1, \xi_2)$ is the cost of moving a unit mass from ξ_1 to ξ_2 on Ξ. Its c-transformation $\phi^c : \xi_2 \rightarrow \mathbb{R}$ is given by

$$
\phi^c(\xi_2) := \inf_{\xi_1 \in \Xi} \{ d(\xi_1, \xi_2) - \phi(\xi_1) \}. \tag{4.13}
$$

Based on the above definition of c-transformation, the Wasserstein distance can be recast as

$$
W_1(P_1, P_2) = \sup_{\phi} \left\{ \int_{\Xi} \phi(\xi_1) P_1(d\xi_1) + \int_{\Xi} \phi^c(\xi_2) P_2(d\xi_2), \forall \xi_1, \xi_2 \in \Xi \right\}. \tag{4.14}
$$

The Kantorovich duality provides an equivalent characterization of the Wasserstein distance. This dual representation demonstrates that the first Wasserstein distance can be calculated by two single integrals with respect to marginal distributions P_1 and P_2, which has much lower computational complexity comparing to directly calculating the double integral under the joint distribution in the original Wasserstein distance formulation. We have the following proposition.

Proposition 4.6 *The DRO problem (4.5), when the Wasserstein distance metric is used, is equivalent to the following convex optimization problem:*

$$
\min_{\mathbf{w}_e, \lambda_1, \lambda_2 \geq 0} \left\{ \lambda_1 \eta_1 + \lambda_2 \eta_2 - \int_{\Xi} \inf_{\xi \in \Xi} \left[\lambda_1 d(\xi, \xi_o) - \frac{1}{2} h(\mathbf{w}_e, \xi) \right] P_o(d\xi_o) \right.
$$
$$
\left. - \int_{\Xi} \inf_{\xi \in \Xi} \left[\lambda_2 d(\xi, \xi_c) - \frac{1}{2} h(\mathbf{w}_e, \xi) \right] P_c(d\xi_c) \right\}. \tag{4.15}
$$

Proof. We first examine the inner maximum problem. From the Lagrangian duality and the minimax inequality, we obtain

$$\sup_{P \in \mathcal{P}} \mathbb{E}_P [h(\mathbf{w}, \boldsymbol{\xi})]$$

$$= \sup_{P \in \mathcal{P}} \inf_{\lambda_1, \lambda_2 \geq 0} \left\{ \int_{\Xi} h(\mathbf{w}, \boldsymbol{\xi}) P(d\boldsymbol{\xi}) + \lambda_1 (\eta_1 - W_1(P, P_o)) + \lambda_2 (\eta_2 - W_1(P, P_c)) \right\} \qquad (4.16)$$

$$\leq \inf_{\lambda_1, \lambda_2 \geq 0} \left\{ \lambda_1 \eta_1 + \lambda_2 \eta_2 + \sup_{P \in \mathcal{P}} \left\{ \int_{\Xi} h(\mathbf{w}, \boldsymbol{\xi}) P(d\boldsymbol{\xi}) - \lambda_1 W_1(P, P_o) - \lambda_2 W_1(P, P_c) \right\} \right\},$$

where λ_1 and λ_2 are Lagrangian multipliers associated with two Wasserstein distance constraints, respectively. For convenience, define

$$a = \sup_{P \in \mathcal{P}} \left\{ \int_{\Xi} h(\mathbf{w}, \boldsymbol{\xi}) P(d\boldsymbol{\xi}) - \lambda_1 W_1(P, P_o) - \lambda_2 W_1(P, P_c) \right\}. \qquad (4.17)$$

We next look into the upper bound on a. It follows from (4.14) that

$$\sup_{P \in \mathcal{P}} \left\{ \int_{\Xi} h(\mathbf{w}, \boldsymbol{\xi}) P(d\boldsymbol{\xi}) - \lambda_1 W_1(P, P_o) - \lambda_2 W_1(P, P_c) \right\}$$

$$= \sup_{P \in \mathcal{P}} \left\{ \int_{\Xi} h(\mathbf{w}, \boldsymbol{\xi}) P(d\boldsymbol{\xi}) - \lambda_1 \sup_{\phi_1} \left\{ \int_{\Xi} \phi_1(\boldsymbol{\xi}) P(d\boldsymbol{\xi}) + \int_{\Xi} \phi_1^c(\boldsymbol{\xi}_o) P_o(d\boldsymbol{\xi}_o), \forall \boldsymbol{\xi}, \boldsymbol{\xi}_o \in \boldsymbol{\xi} \right\} \qquad (4.18)$$

$$- \lambda_2 \sup_{\phi_2} \left\{ \int_{\Xi} \phi_2(\boldsymbol{\xi}) P(d\boldsymbol{\xi}) + \int_{\Xi} \phi_2^c(\boldsymbol{\xi}_c) P_c(d\boldsymbol{\xi}_c), \forall \boldsymbol{\xi}, \boldsymbol{\xi}_c \in \boldsymbol{\xi} \right\} \right\}.$$

Setting $\phi_1(\boldsymbol{\xi}) = \frac{h(\mathbf{w}, \boldsymbol{\xi})}{2\lambda_1}$, $\phi_2(\boldsymbol{\xi}) = \frac{h(\mathbf{w}, \boldsymbol{\xi})}{2\lambda_2}$, we obtain an upper bound of a as follows:

$$\sup_{P \in \mathcal{P}} \left\{ \int_{\Xi} h(\mathbf{w}, \boldsymbol{\xi}) P(d\boldsymbol{\xi}) - \lambda_1 W_1(P, P_o) - \lambda_2 W_1(P, P_c) \right\}$$

$$\leq -\lambda_1 \int_{\Xi} \phi_1^c(\boldsymbol{\xi}_o) P_o(d\boldsymbol{\xi}_o) - \lambda_2 \int_{\Xi} \phi_2^c(\boldsymbol{\xi}_c) P_c(d\boldsymbol{\xi}_c). \qquad (4.19)$$

From Definition 4.5, we have the c-transformation as

$$\phi_1^c(\boldsymbol{\xi}_o) := \inf_{\boldsymbol{\xi} \in \Xi} \{ d(\boldsymbol{\xi}, \boldsymbol{\xi}_o) - \phi_1(\boldsymbol{\xi}) \}, \qquad (4.20)$$

$$\phi_2^c(\boldsymbol{\xi}_c) := \inf_{\boldsymbol{\xi} \in \Xi} \{ d(\boldsymbol{\xi}, \boldsymbol{\xi}_c) - \phi_2(\boldsymbol{\xi}) \}. \qquad (4.21)$$

Combining them with (4.16) and (4.19), we conclude that

$$\sup_{P \in \mathcal{P}} \mathbb{E}_P [h(\mathbf{w}, \boldsymbol{\xi})] \leq \inf_{\lambda_1, \lambda_2 \geq 0} \left\{ \lambda_1 \eta_1 + \lambda_2 \eta_2 - \int_{\Xi} \inf_{\boldsymbol{\xi} \in \Xi} \left[\lambda_1 d(\boldsymbol{\xi}, \boldsymbol{\xi}_o) - \frac{1}{2} h(\mathbf{w}, \boldsymbol{\xi}) \right] P_o(d\boldsymbol{\xi}_o) \right.$$

$$\left. - \int_{\Xi} \inf_{\boldsymbol{\xi} \in \Xi} \left[\lambda_2 d(\boldsymbol{\xi}, \boldsymbol{\xi}_c) - \frac{1}{2} h(\mathbf{w}, \boldsymbol{\xi}) \right] P_c(d\boldsymbol{\xi}_c) \right\}. \qquad (4.22)$$

Algorithm 4.3 : Solving the optimal $\mathbf{w}^*$ for (4.5) using Wasserstein distance-based approach

---

Inputs: Local dataset $\mathcal{S}_{\boldsymbol{\xi}} = \{\boldsymbol{\xi}_1, \boldsymbol{\xi}_2, \ldots, \boldsymbol{\xi}_N\}$, the objective function $h(\mathbf{w}, \boldsymbol{\xi})$, the cloud reference distribution P_c, the edge confidence level $1 - \beta_1$ and the cloud confidence level $1 - \beta_2$.

Outputs: Optimal decision $\mathbf{w}^*$.

1: Compute the empirical distribution $P_o = \frac{1}{N} \sum_{i=1}^{N} \delta_{\boldsymbol{\xi}_i}$.

2: Determine η_1 using $\mathbb{P}\left(W_1(P_t, P_o) \leq \eta_1\right) \geq 1 - \beta_1$.

3: Determine η_2 using $\mathbb{P}\left(W_1(P_t, P_c) \leq \eta_2\right) \geq 1 - \beta_2$.

4: Solve $\min_{\boldsymbol{\xi} \in \boldsymbol{\Xi}} \left[\lambda_1 d(\boldsymbol{\xi}, \boldsymbol{\xi}_o) - \frac{1}{2}h(\mathbf{w}, \boldsymbol{\xi})\right]$ to obtain the optimal function $u^*(\mathbf{w}, \boldsymbol{\xi}_o)$.

5: Solve $\min_{\boldsymbol{\xi} \in \boldsymbol{\Xi}} \left[\lambda_2 d(\boldsymbol{\xi}, \boldsymbol{\xi}_c) - \frac{1}{2}h(\mathbf{w}, \boldsymbol{\xi})\right]$ to obtain the optimal function $v^*(\mathbf{w}, \boldsymbol{\xi}_c)$.

6: Solve (4.15) using the $u^*(\mathbf{w}, \boldsymbol{\xi}_o)$ and $v^*(\mathbf{w}, \boldsymbol{\xi}_c)$ obtained in steps 4 and 5, respectively.

7: Augment $\mathbf{w}^*$ satisfying the optimal value evaluated in step 6.

8: **return** $\mathbf{w}^*$.

We combine the inner worst case result with the outer problem of (4.5), we have the equivalent form given in (4.15). □

It is clear that the above equivalent form (4.15) is a two-layer convex optimization problem, where the two inner problems involve optimization problem over random variable $\boldsymbol{\xi}$ and the outer one is a stochastic program involving the two given distributions P_o and P_c. Thus, the problem (4.15) is amenable to algorithmic implementation. Algorithm 4.3 outlines the steps to obtain the optimal $\mathbf{w}^*$ for the DRO problem (4.5) based on Proposition 4.15. We also note that this equivalent form derived in the Wasserstein approach is more general than that in the KL divergence approach since P_c here is allowed to be arbitrary continuous distributions.

4.3.5   EXPERIMENTAL RESULTS

Since the primary objective of this study is to build a cohesive synergy of the local data and the cloud knowledge for better learning accuracy in IoT applications, we compare the performance of the proposed framework with existing approaches using local edge data only. To ensure fairness, we follow the simulation setups implemented in Abadeh et al. [2015]. Specifically, an underlying true distribution P_t is synthetically generated. The edge information, which is the discrete empirical distribution P_o, is created by sampling N values from P_t. In addition to the aforementioned simulation setups, we further create cloud reference distribution P_c.

In each experiment, we examine the full range of the η_2 to get a clear understanding of the trade-off between the various confidence level corresponding to the cloud knowledge and that of the edge data. In particular, we define the edge-cloud confidence ratio as

$$c = \frac{\eta_1}{\eta_2}. \tag{4.23}$$

Table 4.1: The summary of all the parameters used in logistic regression experiments. $\mathbf{0}$ and $\mathbf{0.1}$ represent the vector of 0's and 0.1's, respectively. $\mathbf{I}_{10}$ is the identity matrix of size 10×10.

Logistic Regression	η_1	w_t	μ_t	Σ_t	w_c	μ_c	Σ_c
(a) $N = 100$	0.02	$[10, 0, 0, 0, 0, 0, 0, 0,$ $0, 0]^T$	$\mathbf{0}$	$\mathbf{I}_{10}$	$[9.5, 0.5, 0, 0, 0, 0, 0,$ $0, 0, 0]^T$	$\mathbf{0.1}$	$\mathbf{I}_{10}$
(b) $N = 200$	0.01						
(c) $N = 300$	0.0067						

In all experiments, η_1 is selected to be a fixed value, and is computed as a function of N with confidence level of 0.95 ($\beta_1 = 0.05$) based on the results discussed in Section 7.3.3. We solve the corresponding optimization problems using the SDPT3 solver in the convex optimization package CVX [Grant et al., 2008] and embedded numerical optimization tools in Matlab.

A Case Study for Logistic Regression. In this experiment, we apply the proposed DRO-based technique to a problem admitting a high-dimensional underlying distribution. In particular, we consider the logistic regression problem as the edge intelligence model and reformulate it using the Wasserstein distance-based DRO framework.

In the logistic regression problem, the underlying distribution maps the 10-dimensional features $\mathbf{x}$ and the corresponding labels y to the probability space. Logistic regression fits a sigmoid function on features and labels to compute the probability of a sample from the feature space belonging to a particular label. The conditional distribution of label y given feature $\mathbf{x}$ is formulated as

$$\text{Prob}(y|\mathbf{x}) = [1 + \exp(-y\langle \mathbf{w}, \mathbf{x}\rangle)]^{-1}. \tag{4.24}$$

Since the conditional distribution defined in (4.24) is a real-valued deterministic function, we utilize the Wasserstein distance which is suitable for quantifying the distance between discrete and continuous distributions. Subsequently, we define the cost function as

$$h(\mathbf{w}, \xi = (\mathbf{x}, y)) = \log(1 + \exp(-y\langle \mathbf{w}, \mathbf{x}\rangle)). \tag{4.25}$$

We also define the distance on the input-output space between two data samples $\xi = (\mathbf{x}, y), \xi' = (\mathbf{x}', y') \in \Xi$ (i.e., the unit cost in the first Wasserstein distance) for classification problems as

$$d(\xi, \xi') = ||\mathbf{x} - \mathbf{x}'|| + |y - y'|, \tag{4.26}$$

where $|| \cdot ||$ represents any norm function.

The objective in this experiment is to accurately predict the logistic regression parameter vector $\mathbf{w}$, given the local data at the edge and the knowledge transfer from the cloud. By plugging

(4.25) in (4.15) and using results from Abadeh et al. [2015], the DRO-based formulation, that incorporates both the cloud and local information, is given by

$$\min_{\mathbf{w}, \lambda_1, \lambda_2, s, t} \lambda_1 \eta_1 + \lambda_2 \eta_2 + \frac{1}{N} \sum_{i=1}^{N} s_i + \int p_c t \, d\mathbf{x} dy, \qquad (4.27)$$

$$s.t. \quad s_i \geq 0.5 h_\mathbf{w}(\boldsymbol{\alpha}_i, +1) - 0.5\lambda_1(1 - \beta_i),$$
$$s_i \geq 0.5 h_\mathbf{w}(\boldsymbol{\alpha}_i, -1) - 0.5\lambda_1(1 + \beta_i),$$
$$t \geq 0.5 h_\mathbf{w}(\mathbf{x}, +1) - 0.5\lambda_2(1 - y),$$
$$t \geq 0.5 h_\mathbf{w}(\mathbf{x}, -1) - 0.5\lambda_2(1 + y),$$
$$\lambda_1 \geq ||\mathbf{w}||_*, \lambda_2 \geq ||\mathbf{w}||_* ..$$

where $||.||_*$ represents the dual norm of its argument. $\boldsymbol{\alpha}_i$ and β_i denote the features of the ith data sample at the network edge and its corresponding label. Note that here the empirical distribution is not characterized by $P_o = \frac{1}{N} \sum_{i=1}^{N} \delta_{\boldsymbol{\xi}=(\boldsymbol{\alpha}_i, \beta_i)}$, where $\delta_{\boldsymbol{\xi}=(\boldsymbol{\alpha}_i, \beta_i)}$ denotes the Dirac point measure at $(\boldsymbol{\alpha}_i, \beta_i)$.

We implement both (4.27) and the one presented in Abadeh et al. [2015] to compare the correct classification rate (CCR) using the parameters in Table 4.1. To this end, we model the underlying feature distribution $P_t(\mathbf{x})$ as a 10-dimensional multivariate standard normal distribution with mean $\boldsymbol{\mu}_t$ and covariance matrix Σ_t. The conditional distribution of labels $P_t(y|\mathbf{x})$ is modeled using (4.24) with sigmoid variable $\mathbf{w}_c$. The joint distribution $P_t(\mathbf{x}, y)$ then can be computed as

$$P_t(\mathbf{x}, y) = P_t(y|\mathbf{x}) P_t(\mathbf{x}). \qquad (4.28)$$

The cloud reference distribution $P_c(\mathbf{x})$ is also modeled as a multivariate normal distribution with mean $\boldsymbol{\mu}_c$ and covariance Σ_c. The conditional distribution of cloud labels is again modeled by (4.24) with parameter $\mathbf{w}_c$ and the joint cloud distribution of feature and label is computed as

$$P_c(\mathbf{x}, y) = P_c(y|\mathbf{x}) P_c(\mathbf{x}). \qquad (4.29)$$

The last integral term in (4.27) is calculated by sample average approximation (SAA) [Kleywegt et al., 2002] method by sampling 1000 points from $P_c(\mathbf{x}, y)$. Figure 4.4 shows the CCR values computed for different scenarios. For computing CCR, we take 10,000 samples from the underlying distribution under each scenario. Figures 4.4a–c illustrate that the CCR improves as the cloud knowledge increases. The improvement in CCR slows down as edge-cloud confidence ratio grows. Figure 4.4d shows the improvement over the formulation presented in Abadeh et al. [2015]. Intuitively, the more empirical samples are collected, the closer the empirical distribution to the underlying distribution is.

A Case Study for Image Classification Using Real Dataset. In order to showcase the efficiency of the proposed framework in more realistic scenarios, we explore our edge intelligence problem

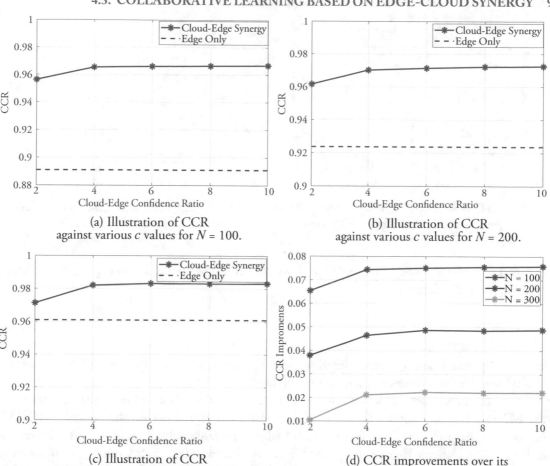

Figure 4.4: Evolution of correct classification ratios against various edge-cloud confidence values and the corresponding CCR improvements achieved in comparison to the technique presented in Abadeh et al. [2015].

on the real dataset MNIST [LeCun et al., 1998], which is widely used for image classification problem, using the Wasserstein distance-based DRO framework. More specifically, we study a convex classification problem with MNIST using softmax regression (i.e., the generalization of logistic regression used for multi-class classfication). The loss function is the cross-entropy error between the predicted and true class. Note that the distance on the input-output space between two data samples is considered as $d(\xi, \xi') = ||\mathbf{x} - \mathbf{x}'|| + \infty \cdot |y - y'|$. To encode cloud information, we first use the complete MNIST dataset to train a conditional generative adversarial networks (cGAN) [Mirza and Osindero, 2014]. Next, we calculate the last integral term

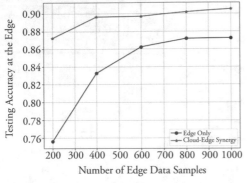

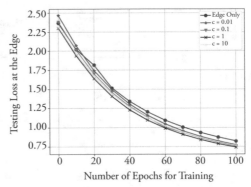

(a) Comparison of Accuracy between DRO Framework and its Edge-Only Counterpart.

(b) Average Testing Loss over Various Cloud-Edge Confidence Values for N = 1,000.

Figure 4.5: Performance evaluation of the Wasserstein distance-based DRO edge leaning framework on MNIST dataset.

in (4.15) by sample average approximation (SAA) [Kleywegt et al., 2002] method by generating 10,000 labeled data using the generator of the trained cGAN.

In the first experiment, we randomly select $N = 200, 400, 600, 800, 1000$ samples from the MNIST training dataset for the edge, and test the performance on the complete MNIST test dataset, respectively. Note that the cloud-edge confidence ratio is fixed to be 1. The results in Figure 4.5a indicate that our proposed DRO edge learning framework outperforms the edge learning using only local dataset in terms of accuracy. The results also illustrates that the performance improvements decrease as the edge local sample number increases, which is consistent with the intuition that the empirical distribution is closer to the underlying distribution as more samples are collected. Next, we set $N = 1000$ and examine the performance of our proposed framework under various cloud-edge confidence ratios in the second experiment. The results in Figure 4.5b show that the average testing loss decreases as the cloud knowledge increases when $c \leq 1$. The improvement over the case of only using edge samples in terms of average testing loss will slow down and even decrease if we further increase the cloud-edge confidence ratio, which is also consistent of our synthetic data simulation results. In a nutshell, by incorporating the cloud knowledge transfer, the proposed DRO-based framework helps the edge node to become more "knowledgeable" about the distribution of the model parameters, and make a more accurate decision with even a small local dataset.

4.4 COLLABORATIVE LEARNING BASED ON KNOWLEDGE TRANSFER OF CONDITIONAL PRIOR DISTRIBUTION

In this section, we turn our attention to the setting where the cloud knowledge transfer is in the form of the conditional prior distribution $P(\mathbf{w}_e|\mathbf{w}_c)$.

4.4.1 CLOUD-EDGE KNOWLEDGE TRANSFER VIA DIRICHLET PROCESS PRIOR

To present the relation between the cloud and the edge, we assume that there are M tasks in the cloud and one edge task, and that all reference k-dimensional model parameters in the cloud, denoted by $\mathcal{W}_c = \{\mathbf{w}_1, \ldots, \mathbf{w}_M\}$, and the edge model parameter $\mathbf{w}_e$ are drawn from a common prior distribution G. The prior distribution G itself is sampled from a Dirichlet process $DP(\alpha, H)$, where $\alpha > 0$ is the concentration parameter and H is the base distribution. The mathematical representation of the Dirichlet process is

$$
\begin{aligned}
\mathbf{w}_m, \mathbf{w}_e &\sim G \\
G &\sim DP(\alpha, H),
\end{aligned}
\tag{4.30}
$$

where $m = 1, \ldots, M$. The advantage of applying the DP prior distribution to hierarchical models has been studied extensively in the statistics literature, see, e.g., Mallick and Walker [1997], Mukhopadhyay and Gelfand [1997], and Müller et al. [2004]. Using the prior distribution sampled from the DP, it is of flexibility to select parameters for different individual learning tasks with high complexity. Further, the DP models the task parameters in a Bayesian nonparametric manner so that we do not need to pre-define a specific form of the common prior.

It follows from the Blackwell–MacQueen urn scheme representation of Dirichlet Process [Teh, 2010], the conditional prior distribution of the edge model $\mathbf{w}_e$ given α, H and cloud model parameters $\mathcal{W}_c = \{\mathbf{w}_1, \ldots, \mathbf{w}_M\}$ is

$$
p(\mathbf{w}_e|\mathcal{W}_c, \alpha, H) = \frac{1}{\alpha + M} \left[\alpha H(\mathbf{w}_e|\nu) + \sum_{m=1}^{M} \delta_{\mathbf{w}_m} \right],
\tag{4.31}
$$

where with a slight abuse of notation, $H(\mathbf{w}_e|\nu)$ denotes the probability density function (PDF) of the base distribution parameterized by ν, and $\delta_{\mathbf{w}_m}$ is the Dirac point measure at $\mathbf{w}_m$. The knowledge transferred from the cloud can be in the form of the above prior distribution of $\mathbf{w}_e$, which is a mixture of the Dirac point measure $\delta_{\mathbf{w}_m}$ corresponding to the meta-knowledge learned from the cloud and the base distribution H corresponding to the base prior of $\mathbf{w}_e$. The edge device uses the prior transferred from each cloud reference model with probability $\frac{1}{\alpha+M}$ and the baseline prior with probability $\frac{\alpha}{\alpha+M}$. The concentration parameter α can be tuned to strike a good balance between the meta-knowledge from the cloud and the base prior. The confidence strength on the cloud increases as α decreases or M increases.

Following Liu et al. [2008], we next use normal distributions to replace the Dirac delta functions:

$$p(\mathbf{w}_e|\mathcal{W}_c, \alpha, H) = \frac{1}{\alpha + M}\left[\alpha H(\mathbf{w}_e|\nu) + \sum_{m=1}^{M} \mathcal{N}\left(\mathbf{w}_e; \mathbf{w}_m, \sigma_m^2 \mathbf{I}\right)\right], \tag{4.32}$$

where $\mathcal{N}\left(\mathbf{w}_e; \mathbf{w}_m, \sigma_m^2 \mathbf{I}\right)$ is a normal distribution with mean $\mathbf{w}_m$ and covariance matrix $\sigma_m^2 \mathbf{I}$. The underlying rationale is that the normal distribution can be regarded as a relaxation of the Dirac point measure. While the Dirac point measure requires the edge model equals to one of the cloud models with probability $\frac{1}{\alpha+M}$, this normal distribution relaxation only requires them to have similar model parameters, which is less stringent in modeling the edge-cloud model relation. Moreover, the normal distribution is more analytically tractable compared to the Dirac delta measure.

4.4.2   DRO FORMULATION WITH CONDITIONAL PRIOR DISTRIBUTION AND EDGE DISTRIBUTION UNCERTAINTY SET

We assume that with large amount of historical data and plenteous computing resources, the cloud data center has the capability to learn $\mathcal{W}_c = \{\mathbf{w}_1, \ldots, \mathbf{w}_M\}$ with enough accuracy. Each model parameter $\mathbf{w}_m$ is learned corresponding to the cloud dataset $\mathcal{D}_c^m = \{(\mathbf{x}_i^m, y_i^m)\}_{i=1}^{N_c^m}$, where $m = 1, \ldots, M$ and $N_c^m \gg N$. We further assume that the prior $p(\mathbf{w}_e|\mathcal{W}_c, \alpha, H)$ is transferred to the edge IoT device, the posterior of $\mathbf{w}_e$ is

$$p(\mathbf{w}_e|\mathcal{D}_e, \mathcal{W}_c, \alpha, H) \propto p(\mathcal{D}_e|\mathbf{w}_e)p(\mathbf{w}_e|\mathcal{W}_c, \alpha, H), \tag{4.33}$$

where the likelihood of the edge data is

$$p(\mathcal{D}_e|\mathbf{w}_e) = \prod_{i=1}^{N} p\left(y_i|\mathbf{x}_i, \mathbf{w}_e\right). \tag{4.34}$$

Then, it remains to characterize the parameter $\mathbf{w}_e$ that maximize the posterior, which is equivalent to minimizing the following negative log-posterior:

$$J\left(\mathbf{w}_e\right) = \frac{1}{N}\sum_{i=1}^{N}\left[-\log p\left(y_i|\mathbf{x}_i, \mathbf{w}_e\right) - \frac{1}{N}\log p(\mathbf{w}_e|\mathcal{W}_c, \alpha, H)\right]. \tag{4.35}$$

For convenience, we define the following real-valued function of $\mathbf{w}_e$ as follows:

$$h(\mathbf{w}_e, \mathbf{x}, y) = -\log p\left(y|\mathbf{x}, \mathbf{w}_e\right) - \frac{1}{N}\log p(\mathbf{w}_e|\mathcal{W}_c, \alpha, H). \tag{4.36}$$

The negative log-posterior (4.35) then can be reformulated as an expectation of $h(\mathbf{w}_e, \mathbf{x}, y)$ over the empirical distribution of the edge data sample set $\mathcal{S}_e$:

$$J\left(\mathbf{w}_e\right) = \mathbb{E}_{(\mathbf{x}, y) \sim P_o}[h(\mathbf{w}_e, \mathbf{x}, y)]. \tag{4.37}$$

Thus, the edge learning with the cloud knowledge can be presented as the following formulation:

$$\min_{\mathbf{w}_e} \mathbb{E}_{(\mathbf{x},y) \sim P_o}[h(\mathbf{w}_e, \mathbf{x}, y)]. \tag{4.38}$$

Considering the limited computing resources and amount of the local data at the edge node, we cast the above edge learning problem as a DRO problem to achieve robust decision making, outlined as follows:

$$\textbf{P1:} \min_{\mathbf{w}_e} \max_{P} \mathbb{E}_{(\mathbf{x},y) \sim P}[h(\mathbf{w}_e, \mathbf{x}, y)], \tag{4.39}$$

$$s.t. \ D(P, P_o) \leq \eta,$$

where $D(P, P_o)$ is a distribution distance function of P and P_o. The above distribution distance based uncertainty set is constructed in a way that the distribution of $(\mathbf{x}, y)$ are contained in a distributional ball centered around the local empirical distribution P_o with radius η. With proper selection of η, the uncertainties at the edge can be reduced such that the uncertainty set could be obtained for a better estimation of the underlying distribution P_t.

4.4.3 A DUALITY APPROACH TO THE INNER MAXIMIZATION PROBLEM

We solve **P1** by reducing it to a single-layer optimization problem via a duality approach. Plugging (4.36) into the inner maximization problem of **P1** and using the first-order Wasserstein Distance $W_1(P, P_o)$ as $D(P, P_o)$ in the space of probability distributions, we can have

$$\min_{\mathbf{w}_e} \left\{ -\frac{1}{N} \log p(\mathbf{w}_e | \mathcal{W}_c, \alpha, H) + \max_{P} \mathbb{E}_{(\mathbf{x},y) \sim P}[-\log p(y|\mathbf{x}, \mathbf{w}_e)] \right\}, \tag{4.40}$$

$$s.t. \ W_1(P, P_o) \leq \eta.$$

Next, we define a distance function on the input-output space for classification problems, which will be needed later.

Definition 4.7   Distance function on the input-output space for classification.   The distance on the input-output space between two data samples $\boldsymbol{\xi} = (\mathbf{x}, y), \boldsymbol{\xi}' = (\mathbf{x}', y') \in \Xi$ for classification problems is defined as follows:

$$d(\boldsymbol{\xi}, \boldsymbol{\xi}') = ||\mathbf{x} - \mathbf{x}'|| + \beta|y - y'|, \tag{4.41}$$

where $|| \cdot ||$ represents a norm function and β is a positive constant.

We first study the inner maximization problem of (4.40), for which we impose the following assumption.

Assumption 4.8   [Shafieezadeh-Abadeh et al., 2017] The negative logarithmic function of parametric probabilistic model $-\log p(y|\mathbf{x}, \mathbf{w}_e)$ can be re-expressed as a convex and Lipschitz continuous loss function $L(z)$, where $z = y\langle \mathbf{w}_e, \mathbf{x} \rangle$ if $p(y|\mathbf{x}, \mathbf{w}_e)$ is a classification model (and $z = \langle \mathbf{w}_e, \mathbf{x} \rangle - y$ if $p(y|\mathbf{x}, \mathbf{w}_e)$ is a regression model).

We also need the following lemma, which states that the worst-case expectation over a Wasserstein distance ball can be treated as a classical robust optimization problem using the duality approach.

Lemma 4.9 Strong duality [Gao and Kleywegt, 2016]   For any measurable loss function $l(\boldsymbol{\xi})$ we have

$$\sup_{W_1(P,P_o)\leq\eta} \mathbb{E}_{\boldsymbol{\xi}\sim P}[l(\boldsymbol{\xi})] = \inf_{\lambda\geq 0} \left\{ \lambda\eta + \frac{1}{N}\sum_{i=1}^{N} \sup_{\boldsymbol{\xi}\in\Xi} [l(\boldsymbol{\xi}) - \lambda d(\boldsymbol{\xi}, \boldsymbol{\xi}_i)] \right\}, \qquad (4.42)$$

where $\boldsymbol{\xi}_i = (\mathbf{x}_i, y_i)$ is the ith data sample and λ is the Lagrangian multiplier.

Applying Lemma 4.9 to the inner maximization problem of the original problem (4.40) and combining it with the outer minimization problem, we obtain that

$$\mathbf{P2} \min_{\mathbf{w}_e,\lambda} \left\{ -\frac{1}{N}\log p(\mathbf{w}_e|\mathcal{W}_c, \alpha, H) + \frac{1}{N}\sum_{i=1}^{N} s_i(\mathbf{w}_e) + \lambda\eta \right\}, \qquad (4.43)$$

$$s.t. \ \lambda \geq 0,$$
$$s_i(\mathbf{w}_e) = \sup_{(\mathbf{x},y)\in\Xi} [-\log p(y|\mathbf{x}, \mathbf{w}_e) - \lambda d((\mathbf{x}, y), (\mathbf{x}_i, y_i))].$$

In what follows, we recast **P2** as a single-layer optimization problem. We first impose the following lemma.

Lemma 4.10 Supremum representation [Shafieezadeh-Abadeh et al., 2017]   Suppose $L(z)$ is a convex and Lipschitz continuous loss function, $\boldsymbol{\theta}, \boldsymbol{\zeta}' \in \mathbb{R}^d$ and $\gamma > 0$. If $lip(L)\|\boldsymbol{\theta}\|_* \leq \gamma$, then we have the following result:

$$\sup_{\boldsymbol{\zeta}\in\mathbb{R}^d} L(\langle\boldsymbol{\theta},\boldsymbol{\zeta}\rangle) - \gamma\|\boldsymbol{\zeta} - \boldsymbol{\zeta}'\| = L(\langle\boldsymbol{\theta},\boldsymbol{\zeta}'\rangle), \qquad (4.44)$$

where $lip(L)$ is the Lipschitz modulus of L and $\|\cdot\|_*$ is the dual form of arbitrary dual function $\|\cdot\|$.

We have the following propositions for classification and regression edge learning models respectively by using the result in Lemma 4.10.

Proposition 4.11   *Suppose Assumption 4.8 holds. When the probabilistic model $P(y|\mathbf{x}, \mathbf{w}_e)$ is for classification, the DRO problem (4.40) can be reduced to the following single-layer optimization prob-*

lem:

$$\min_{\mathbf{w}_e, \lambda} \left\{ -\frac{1}{N} \log p(\mathbf{w}_e | \mathcal{W}_c, \alpha, H) + \frac{1}{N} \sum_{i=1}^{N} s_i(\mathbf{w}_e) + \lambda \eta \right\}, \tag{4.45}$$

$$\text{s.t. } \lambda \geq lip(L) \|\mathbf{w}_e\|_*,$$
$$s_i(\mathbf{w}_e) = \max \{ -\log p(1|\mathbf{x}_i, \mathbf{w}_e) - \lambda\beta|1 - y_i|,$$
$$-\log p(-1|\mathbf{x}_i, \mathbf{w}_e) - \lambda\beta| - 1 - y_i| \}.$$

Proof. We first plug (4.41) into $s_i(\mathbf{w}_e)$, and obtain the following:

$$s_i(\mathbf{w}_e) = \sup_{(\mathbf{x}, y) \in \Xi} [-\log p(y|\mathbf{x}, \mathbf{w}_e) - \lambda \|\mathbf{x} - \mathbf{x}_i\| - \lambda\beta|y - y_i|]$$

$$= \max \left\{ \sup_{\mathbf{x} \in \mathbb{R}^u} [-\log p(1|\mathbf{x}, \mathbf{w}_e) - \lambda \|\mathbf{x} - \mathbf{x}_i\| - \lambda\beta|1 - y_i|], \right.$$
$$\left. \sup_{\mathbf{x} \in \mathbb{R}^u} [-\log p(-1|\mathbf{x}, \mathbf{w}_e) - \lambda \|\mathbf{x} - \mathbf{x}_i\| - \lambda\beta| - 1 - y_i|] \right\}. \tag{4.46}$$

According to Lemma 4.10 and Assumption 4.8, the above supremum can be further simplified as follows:

$$s_i(\mathbf{w}_e) = \max \{ -\log p(1|\mathbf{x}_i, \mathbf{w}_e) - \lambda\beta|1 - y_i|, -\log p(-1|\mathbf{x}_i, \mathbf{w}_e) - \lambda\beta| - 1 - y_i| \}, \tag{4.47}$$

where $\lambda \geq lip(L) \|\mathbf{w}_e\|_*$.

Plugging (4.47) into **P2,** we obtain (4.45), as the equivalent form of the original problem (4.40). $\qquad \square$

Proposition 4.12   *Suppose Assumption 4.8 holds. When the probabilistic model $P(y|\mathbf{x}, \mathbf{w}_e)$ is for regression, the DRO problem (4.40) can be reduced to the following single-layer optimization problem:*

$$\min_{\mathbf{w}_e} \left\{ -\frac{1}{N} \log p(\mathbf{w}_e | \mathcal{W}_c, \alpha, H + \frac{1}{N} \sum_{i=1}^{N} -\log p(y_i|\mathbf{x}_i, \mathbf{w}_e) + \eta \, lip(L) \|(\mathbf{w}_e, -1)\| \right\}. \tag{4.48}$$

Proof. Suppose Assumption 4.8 holds, we first look into $s_i(\mathbf{w}_e)$ in **P2.** We applied Lemma 4.10 to it to obtain

$$s_i(\mathbf{w}_e) = \sup_{(\mathbf{x}, y) \in \Xi} [-\log p(y|\mathbf{x}, \mathbf{w}_e) - \lambda d((\mathbf{x}, y), (\mathbf{x}_i, y_i))]$$

$$= \sup_{(\mathbf{x}, y) \in \Xi} [L(\langle \mathbf{w}_e, \mathbf{x} \rangle - y) - \lambda \|(\mathbf{x}, y) - (\mathbf{x}_i, y_i)\|]$$

$$= L(\langle \mathbf{w}_e, \mathbf{x}_i \rangle - y_i), \tag{4.49}$$

where $\lambda \geq lip(L)||(\mathbf{w}_e, -1)||$.

Plugging (4.49) into **P2**, we obtain

$$\min_{\mathbf{w}_e, \lambda} \left\{ -\frac{1}{N} \log p(\mathbf{w}_e | \mathcal{W}_c, \alpha, H) + \frac{1}{N} \sum_{i=1}^{N} -\log p\left(y_i | \mathbf{x}_i, \mathbf{w}_e\right) + \eta \lambda \right\} \tag{4.50}$$
$$s.t. \ \lambda \geq lip(L)||(\mathbf{w}_e, -1)||_*,$$

which can be further represented as (4.48). □

4.4.4   A CONVEX RELAXATION FOR THE OUTER MINIMIZATION PROBLEM

Though we have reduced the original DRO problem to a single-layer optimization problem, the first term in (4.45) and (4.48) makes this single-layer optimization problem nonconvex with respect to $\mathbf{w}_e$ since the first term is a negative logarithm of a PDF mixture, which is not convex. Hence, it is not practical to directly use numerical methods to solve (4.45) and (4.48). In what follows, we propose an Expectation-Maximization (EM) algorithm-inspired method to provide a convex relaxation with respect to $\mathbf{w}_e$ for the outer minimization problem, and devise the learning algorithm accordingly.

We first define a new distribution $Q(z)$ with $\sum_{m=0}^{M} Q(z_m) = 1$ and $Q(z_m) \geq 0$. We further define the σ_m-norm of $\mathbf{x}$ as $||\mathbf{x}||_{\sigma_m}^2 = \mathbf{x}^T \sigma_m^2 \mathbf{I}^{-1} \mathbf{x}$. We can rewrite the optimization problem **P2** as

$$\min_{\mathbf{w}_e, \lambda \geq 0} \left\{ \frac{1}{N} \sum_{i=1}^{N} s_i(\mathbf{w}_e) + \eta\lambda - \frac{1}{N} \log \left[\frac{\alpha}{\alpha + M} Q(z_0) \frac{H(\mathbf{w}_e | v)}{Q(z_0)} \right. \right.$$
$$\left. \left. + \frac{1}{\alpha + M} \sum_{m=1}^{M} \frac{Q(z_m)}{\sigma_m \sqrt{(2\pi)^k}} \frac{\exp(-0.5||\mathbf{w}_e - \mathbf{w}_m||_{\sigma_m}^2)}{Q(z_m)} \right] \right\}. \tag{4.51}$$

For convenience, we next define

$$a(\mathbf{w}_e) = -\frac{1}{N} \log \left[\frac{\alpha}{\alpha + M} Q(z_0) \frac{H(\mathbf{w}_e | v)}{Q(z_0)} \right.$$
$$\left. + \frac{1}{\alpha + M} \sum_{m=1}^{M} \frac{Q(z_m)}{\sigma_m \sqrt{(2\pi)^k}} \frac{\exp(-0.5||\mathbf{w}_e - \mathbf{w}_m||_{\sigma_m}^2)}{Q(z_m)} \right], \tag{4.52}$$

$$b(\mathbf{w}_e) = \frac{1}{N} \sum_{i=1}^{N} s_i(\mathbf{w}_e) + \eta\lambda. \tag{4.53}$$

Algorithm 4.4 : Solving the optimal $\mathbf{w}_e^*$ for (4.40)

Inputs: Local dataset $\mathcal{D}_e = \{\xi_i = (\mathbf{x}_i, y_i)\}_{i=1}^N$, the radius η, the concentration parameter α, the base distribution H, the learning rate ϵ and M cloud reference classifier parameters $\mathcal{W}_c = \{\mathbf{w}_1, \ldots, \mathbf{w}_M\}$.

Outputs: Optimal decision $\mathbf{w}_e^*$.

1: Initialize $\mathbf{w}_e^{(0)}$, $\lambda^{(0)}$ and let $t = 1$.
2: Compute $\gamma_m(\mathbf{w}_e)$ the empirical distribution $P_o = \frac{1}{N} \sum_{i=1}^N \delta_{\xi_i}$.
3: Compute the variables $Q^{(t)}(z_0)$ and $Q^{(t)}(z_m)$ using (4.55) and (4.56), respectively.
4: Gradient descending for $\mathbf{w}_e$: compute $\mathbf{w}_e^{(t)} = \mathbf{w}_e^{(t-1)} - \epsilon \nabla_{\mathbf{w}_e}[l(\mathbf{w}_e|\mathbf{w}_e^{(t-1)})]$.
5: Gradient descending for λ: compute $\lambda^{(t)} = \lambda^{(t-1)} - \epsilon \nabla_\lambda[l(\mathbf{w}_e|\mathbf{w}_e^{(t-1)})]$.
6: If $a(\mathbf{w}_e^{(t)}) + b(\mathbf{w}_e^{(t)})$ converges, the algorithm stops; otherwise, go back to 3.
7: **return** $\mathbf{w}_e^*$.

Since the logarithmic function is concave, we obtain the upper bound of $a(\mathbf{w}_e)$ by applying Jensen's Inequality as follows:

$$a(\mathbf{w}_e) \leq -\frac{1}{N} \left\{ Q(z_0) \log \left[\frac{\alpha}{\alpha + M} \frac{H(\mathbf{w}_e|\nu)}{Q(z_0)} \right] \right. $$
$$\left. + \sum_{m=1}^M Q(z_m) \log \left[\frac{1}{\alpha + M} \frac{\exp(-0.5\|\mathbf{w}_e - \mathbf{w}_m\|_{\sigma_m}^2)}{\sigma_m \sqrt{(2\pi)^k} Q(z_m)} \right] \right\}. \tag{4.54}$$

Consider the learning process at iteration t for the outer minimization problem. Given $\mathbf{w}_e^{(t-1)}$ from the iteration $t - 1$, the upper bound on $a(\mathbf{w}_e^{(t-1)})$ can be achieved with the following $\{Q^{(t)}(z_m), m = 0. \ldots, M\}$ determined in the E-step:

(E-step)

$$Q^{(t)}(z_0) = \frac{\alpha H(\mathbf{w}_e^{(t-1)}|\nu)}{\sum_{m=1}^M \frac{\exp(-0.5\|\mathbf{w}_e^{(t-1)} - \mathbf{w}_m\|_{\sigma_m}^2)}{\sigma_m \sqrt{(2\pi)^k}} + \alpha H(\mathbf{w}_e^{(t-1)}|\nu)}, \tag{4.55}$$

$$Q^{(t)}(z_m) = \frac{\exp(-0.5\|\mathbf{w}_e^{(t-1)} - \mathbf{w}_m\|_{\sigma_m}^2)/(\sigma_m \sqrt{(2\pi)^k})}{\sum_{m=1}^M \frac{\exp(-0.5\|\mathbf{w}_e^{(t-1)} - \mathbf{w}_m\|_{\sigma_m}^2)}{\sigma_m \sqrt{(2\pi)^k}} + \alpha H(\mathbf{w}_e^{(t-1)}|\nu)}. \tag{4.56}$$

Next, define

$$
\begin{aligned}
l(\mathbf{w}_e | \mathbf{w}_e^{(t-1)}) = -\frac{1}{N} \Bigg\{ & Q^{(t)}(z_0) \log \left[\frac{\alpha}{\alpha + M} \frac{H(\mathbf{w}_e | \nu)}{Q^{(t)}(z_0)} \right] \\
& + \sum_{m=1}^{M} Q^{(t)}(z_m) \log \left[\frac{1}{\alpha + M} \frac{\exp(-0.5 \|\mathbf{w}_e - \mathbf{w}_m\|_{\sigma_m}^2)}{\sigma_m \sqrt{(2\pi)^k} Q^{(t)}(z_m)} \right] \Bigg\} + \frac{1}{N} \sum_{i=1}^{N} s_i(\mathbf{w}_e) + \eta \lambda,
\end{aligned}
\tag{4.57}
$$

based on which the edge model parameters are updated in the M-step by minimizing (4.57):

$$
\textbf{(M-step)} \qquad \mathbf{w}_e^{(t)} := \arg\min_{\mathbf{w}_e} l(\mathbf{w}_e | \mathbf{w}_e^{(t-1)}).
\tag{4.58}
$$

To illustrate the convergence behavior of the proposed approach, we combine (4.54)–(4.57) and conclude that

$$
a(\mathbf{w}_e^{(t)}) + b(\mathbf{w}_e^{(t)}) \leq l(\mathbf{w}_e^{(t)} | \mathbf{w}_e^{(t-1)}),
\tag{4.59}
$$
$$
a(\mathbf{w}_e^{(t-1)}) + b(\mathbf{w}_e^{(t-1)}) = l(\mathbf{w}_e^{(t-1)} | \mathbf{w}_e^{(t-1)}).
\tag{4.60}
$$

Intuitively, with an improved $\mathbf{w}_e^{(t)}$ compared to the previous $\mathbf{w}_e^{(t-1)}$, i.e.,

$$
l(\mathbf{w}_e^{(t)} | \mathbf{w}_e^{(t-1)}) \leq l(\mathbf{w}_e^{(t-1)} | \mathbf{w}_e^{(t-1)}),
\tag{4.61}
$$

it is guaranteed that

$$
a(\mathbf{w}_e^{(t)}) + b(\mathbf{w}_e^{(t)}) \leq a(\mathbf{w}_e^{(t-1)}) + b(\mathbf{w}_e^{(t-1)}),
\tag{4.62}
$$

which shows that the above EM-inspired approach monotonically improves minimizing **P2**.

The algorithmic implementation of (4.61) can be accomplished by many numerical optimization methods. For instance, the gradient descent algorithm can be applied iteratively to (4.61) to produce successively improved versions of $\mathbf{w}_e$ until convergence. This method is guaranteed to converge to one of local minimas or saddle points of the original problem **P2** since $l(\mathbf{w}_e | \mathbf{w}_e^{(t-1)})$ is continuous with respect to both of $\mathbf{w}_e^{(t-1)}$ and $\mathbf{w}_e$ [Wu et al., 1983]. Algorithm 4.4 outlined steps of obtaining the optimal $\mathbf{w}_e$ for the DRO problem (4.40) based on Propositions 4.11 and 4.12 and the above convex relaxation method. To mitigate the issue of convergence to saddle points, several initial points $\mathbf{w}_e$ can be randomly selected at step 1 of Algorithm 4.3 [Wu et al., 1983].

4.4.5   EXPERIMENTAL RESULTS

In this section, we present the experimental setup and then evaluate the proposed edge learning algorithm for both of synthetic and real-world datasets. In particular, we first quantify the corresponding performance gain over the standard approach using local edge data only. We then investigate the impact of the parameter α, which is the concentration parameter of the relaxed

DP, on the performance of our edge learning algorithm. The value of α controls the confidence strength corresponding to the cloud knowledge and that of the edge local data. Since it is highly nontrivial to analyze the impact of α on the proposed algorithm, we here provide some experimental exploration on this issue. When α increases, the knowledge transfer process from the cloud to the edge whittles down since the proposed algorithm puts more confidence strength on the edge local data for the edge leaning; when α decreases, the knowledge transfer process becomes stronger since the proposed algorithm puts more confidence strength on the cloud information. With proper choices of α, we expect that the proposed algorithm lies in the middle and provides a balanced trade-off between the cloud and the edge to induce an appropriate knowledge transfer. Note that the mean and variance of the base distribution H are specified as $\mu = \mathbf{0}$ and $\Sigma = \mathbf{I}$, and σ_m^2 is set to be 1 in the meta-knowledge from the cloud in all the following experiments.

To simplify the experiment setting, instead of implementing EM algorithm-inspired method from the scratch, we approximately set $Q^{(t)}(z_0) = \frac{\alpha}{M+\alpha}$ and $Q^{(t)}(z_0) = \frac{1}{M+\alpha}$, the loss function (4.57) can reduce to:

$$l(\mathbf{w}_e) = -\frac{1}{N}\left\{\frac{\alpha}{\alpha+M}\log\left[H(\mathbf{w}_e|\nu)\right] - \sum_{m=1}^{M}\frac{\alpha}{2(\alpha+M)}||\mathbf{w}_e - \mathbf{w}_m||_{\sigma_m}^2\right) $$

$$+ \frac{1}{N}\sum_{i=1}^{N}s_i(\mathbf{w}_e) + \eta\lambda, \tag{4.63}$$

which is much less computational demanding and improves the deployability of our algorithm on extremely resource-limited edge devices.

Classification on Synthetic Datasets. In this experiment, we consider a binary classification problem as the edge learning objective. In particular, we use the logistic regression model as the edge learning model, which maps the feature $\mathbf{x}$ and the corresponding label y to the probability space. A sigmoid function is fitted on features and labels to compute the probability of a sample from the feature space belonging to a particular label. The parametric probabilistic model of label y given feature $\mathbf{x}$ is formulated as

$$P(y|\mathbf{x}, \mathbf{w}_e) = [1 + \exp(-y\langle\mathbf{w}_e, \mathbf{x}\rangle)]^{-1}. \tag{4.64}$$

To evaluate the impact of DP concentration parameter α on the performance of the proposed algorithm for the classification problem, we follow a similar setup implemented in Liu et al. [2008] to generate two different synthetic datasets for the cloud and the edge, respectively. In each synthetic dataset, the true binary class labeled data from each class are generated from a Gaussian mixture model (GMM). Specifically, for the cloud dataset, the class 1 data samples are drawn from a three-component GMM defined as follows. Mixture weights are (three components) (0.3, 0.3, 0.4); respective 2D means are (1, 1), (3, 3), and (5, 5); and respective covariance matrices are $\Sigma_1 = \begin{pmatrix} 0.3 & 0.7 \\ 0.7 & 3.0 \end{pmatrix}$, $\Sigma_2 = \begin{pmatrix} 3.0 & 0.0 \\ 0.0 & 0.3 \end{pmatrix}$, and $\Sigma_3 = \begin{pmatrix} 3.0 & -0.5 \\ -0.5 & 0.3 \end{pmatrix}$.

Samples for class 2 in the cloud dataset are drawn from a single 2D Gaussian distribution with mean $(2.5, 1.5)$ and diagonal covariance with symmetric variance 0.5. For edge dataset, the class 1 data samples are drawn from a three-component GMM defined as follows. Mixture weights are (three components) $(0.3, 0.3, 0.4)$; the respective 2D means are $(0.9, 0.9)$, $(3, 3)$, and $(5, 5)$; and the respective covariance are Σ_1, Σ_2, and Σ_3. The data for class 2 in the edge dataset are drawn from a single 2D Gaussian distribution with mean $(2.4, 1.8)$, and diagonal covariance with symmetric variance 0.5. The metric function on the feature-label space is considered as $d(\xi, \xi') = ||\mathbf{x} - \mathbf{x}'|| + \infty \cdot |y - y'|$.

We first obtain cloud reference logistic regression model parameters $(\mathbf{w}_1, \ldots, \mathbf{w}_5)$ via training with 10,000 data samples from the above cloud dataset for 5 times. We then set different concentration parameters $\alpha \in \{0.1, 1, 10, 100, 1000, 10,000\}$ and train the edge model parameters $\mathbf{w}_e$ with $N \in \{20, 40, 60, 80\}$ samples from the edge dataset respectively to get a clear understanding of the trade-off between the cloud knowledge and the edge data. Note that each value of the radius is set to its minimal value η^* with confidence level of 0.95, which can be computed as a function of N based on (4.11). The testing accuracy of the classifier is evaluated with 1000 testing samples from the edge dataset. Each data point of testing results is obtained by averaging over 10 independent trials. Given the ground truth that the edge dataset is closely related to the cloud one, we can observe that our proposed Dirichlet process prior-based DRO algorithm efficiently transfer the knowledge since the proposed approach outperforms the standard approach using local edge data only from Figure 4.6. Figure 4.6a illustrates that the learning performance of our proposed algorithm will be slightly improved by using more samples at the edge. The testing accuracy has almost no change when one tries to further increase sample number N from 60–80, which is also consistent to the observation that the performance of only using the local dataset is approaching the performance of our proposed algorithm when $N = 80$. These observations can be explained as the edge has enough confidence to learn an accurate model based on its own local samples without accepting the transferred knowledge from the cloud. Figure 4.6b illustrates that the edge learning performance can be improved by decreasing the value of concentration parameter α when data samples are limited. When the sample number is increased from 40–60 in Figure 4.6c, the optimal testing accuracy is obtained roughly at $\alpha = 1$, which is a balanced trade-off point providing the appropriate knowledge transfer from cloud to edge.

Regression on Synthetic Datasets. We next evaluate our proposed Dirichlet process prior-based DRO algorithm on synthetic datasets for regression problems. Specifically, we consider a linear regression problem on one dimensional data. Following the same intuition of classification experiments, two different training datasets are generated for the cloud and the edge, respectively. The synthetic sample (x_c, y_c) from cloud datasets is generated from the underlying linear model $y_c = 3x_c + 2 + 0.2\epsilon$ where $x_c \sim \mathcal{N}(0, 1)$ and $\epsilon \sim \mathcal{N}(0, 1)$. The edge dataset

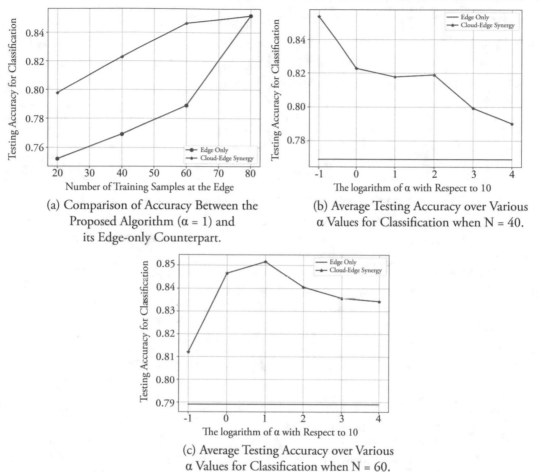

Figure 4.6: Performance evaluation of the proposed Dirichlet process prior-based DRO algorithm for classification on synthetic dataset.

sample (x_e, y_e) is generated from another underlying linear model $y_e = 3x_e + 2 + \epsilon$ where $x_c \sim \mathcal{N}(0, 1)$ and $\epsilon \sim \mathcal{N}(0, 1)$. The objective is to fit the model with the mean squared error (MSE) loss function.

Similar to the classification case, we start with obtaining cloud reference linear regression model parameters $(\mathbf{w}_1, \ldots, \mathbf{w}_3)$ by training with 1000 data samples generated from the cloud underlying linear model for 3 times. We then use limited samples from the edge dataset and cloud reference model parameters to train the edge parameters $\mathbf{w}_e$ with various values of concentration parameter α and the radius η. We examine the out-of-samples performance of the proposed algorithm by evaluating the testing loss with 1000 testing samples from the slightly

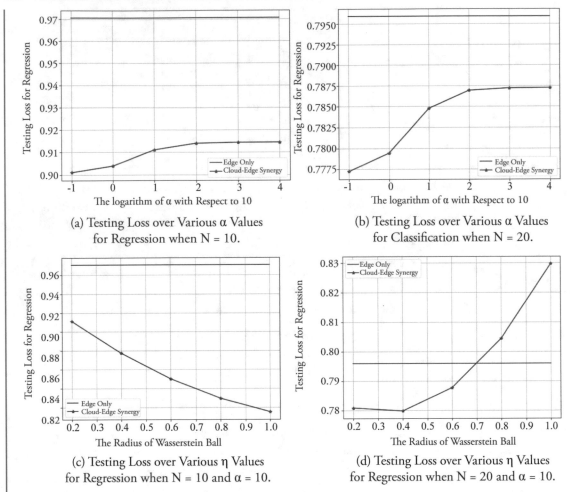

(a) Testing Loss over Various α Values
for Regression when N = 10.

(b) Testing Loss over Various α Values
for Classification when N = 20.

(c) Testing Loss over Various η Values
for Regression when N = 10 and α = 10.

(d) Testing Loss over Various η Values
for Regression when N = 20 and α = 10.

Figure 4.7: Performance evaluation of the proposed Dirichlet process prior-based DRO algorithm for regression on synthetic dataset.

shifted underlying linear model $y_e = 3x_e + 2 + 0.8\epsilon$ where $x_c \sim \mathcal{N}(0, 1)$ and $\epsilon \sim \mathcal{N}(0, 1)$ at the edge.

We compare the performance of our proposed algorithm with the case only using local dataset at the edge. From Figures 4.7a,b, we can observe that the proposed approach outperforms the standard approach using local edge data only. Moreover, the performance is improved by increasing the value of α since the edge put more confidence on the knowledge from the cloud. Unsurprisingly, Figure 4.7c illustrates that the performance will be further improved if we increase the value of the radius η. This is reasonable because the Wasserstein ball can contain the true underlying distribution of the edge dataset with higher probability (i.e., larger than 0.95) when we increases η. The performance drops if we further increase the value of the radius η when the sample number is increased from 10–20, which can be observed from Figure 4.7d. The reason may be that, when we introduce more uncertainty by increasing η, the performance of our algorithm will be in turn hurt since the edge has enough local samples to learn an accurate model.

Image Classification Based on MNIST Dataset. In order to showcase the efficiency of the proposed algorithm in more realistic scenarios, we further explore our edge intelligence problem on the real dataset MNIST [LeCun et al., 1998], which is widely used for image classification problems, using the Wasserstein distance-based DRO algorithm. More specifically, we study a convex classification problem with MNIST using softmax regression (i.e., the generalization of logistic regression used for multi-class classfication). The loss function is the cross-entropy error between the predicted and true class. Note that the distance on the input-output space between two data samples is considered as $d(\xi, \xi') = ||\mathbf{x} - \mathbf{x}'|| + \infty \cdot |y - y'|$. To encode cloud information, we first use the complete MNIST dataset to train a cGAN [Mirza and Osindero, 2014]. We then used the trained generator of cGAN to generate 10,000 fake image samples, which can be considered as a distributionally shifted cloud dataset compared to ordinary MNIST one.

We first use the generated cloud dataset to obtain cloud reference model parameters $(\mathbf{w}_1, \ldots, \mathbf{w}_3)$ via training the classifier for three times. We then randomly select $N = 100, 200, \ldots, 800$ samples from the ordinary MNIST training dataset for the edge training samples, and test the performance on the complete MNIST test dataset, respectively. The results in Figure 4.8a indicate that our proposed DRO edge learning algorithm outperforms the edge learning approach using only local dataset in terms of accuracy. The results also illustrate that the performance improvements decrease as the edge local sample number increases, which is consistent with the truth that the empirical distribution is closer to the underlying distribution as more samples are collected. Next, we set $N = 200$ and examine the performance of our proposed algorithm under various values of the concentration parameter α. The results in Figure 4.8b show that the average testing accuracy increases as the confidence level of cloud knowledge increases (i.e., decreasing the value of α).

4.5 SUMMARY

In this chapter, we propose a DRO-based edge intelligence approach to achieve the real-time edge intelligence based on the innovative synergy of the local data processing and the knowledge transferred from the cloud. Because of the modeling power of DRO for decision making under uncertainty, we resort to a refined DRO formulation to deal with the limitation of computing resources and local data amount at the network edge. In particular, we investigate two interesting uncertainty models corresponding to the cloud knowledge transfer, one in terms of the distribution uncertainty set based on the cloud data distribution, and the other in terms of the prior

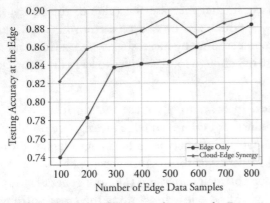

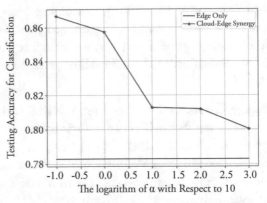

(a) Comparison of Accuracy between the Proposed Algorithm and its Edge-Only Counterpart.

(b) Testing Accuracy over Various α Values for MNIST Classification when N = 200.

Figure 4.8: Performance evaluation of the proposed Dirichlet process prior-based DRO algorithm on MNIST dataset.

distribution of the edge model conditioned on the cloud model. Furthermore, we transform the original min-max optimization problem to an equivalent convex optimization problem. Experimental results demonstrate that the proposed edge leaning algorithm outperforms the standard approach using local edge data only.

CHAPTER 5

Hierarchical Mobile-Edge-Cloud Model Training with Hybrid Parallelism

In this chapter, we present an execution paradigm of hybrid parallelism to accelerate the DNN model training process under the hierarchical mobile-edge-cloud architecture.

5.1 INTRODUCTION

Recall that in Chapter 4, we study how to collaboratively train an EI model via edge-edge collaboration or edge-cloud collaboration. In this chapter, we move forward to study how to train an EI model without the assumption of a pre-trained model in the cloud. For this problem, a natural idea is to train the DNN in a fully decentralized peer-to-peer manner at the edge side [Mathur and Chahal, 2018]. This approach avoids the communication overhead between the edge devices and the cloud center. However, when the computation resources of the edge devices are limited, solely relying on them to train the DNN is impractical, or may cause significant computation delay. We classify this approach and the cloud training approach as *horizontal training*, as the computation tasks are executed over multiple workers at the same system level (either the computing units in the cloud, or the edge devices in the fully distributed peer-to-peer network).

Alternatively, it is plausible to take *hierarchical training* approaches to efficient training of DNNs. For example, JointDNN [Eshratifar et al., 2018] is proposed to train some layers of a DNN on an edge device and the other layers on the cloud center. However, the latency between the edge device and the cloud center is still one major hurdle limiting the training speed. The emerging edge computing paradigm provides another option in which the edge servers are in between the edge devices and the cloud center, and can execute the computation tasks as close as possible to the data sources. Compared to that between the cloud and the edge devices, the communication latency between the edge servers and edge devices is much lower. These excellent properties motivate the emerging edge learning scheme of jointly training a DNN with an edge device and an edge server [Ren et al., 2019]. Figure 5.1 illustrates the difference between the

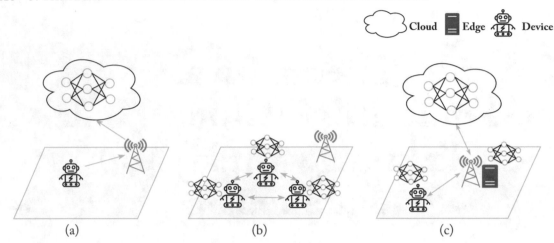

Figure 5.1: Training a DNN: (a) on the cloud center; (b) in the edge devices in a fully decentralized peer-to-peer manner; and (c) on the mobile-edge-cloud hierarchical architecture. Here (a) and (b) belong to *horizontal training*, while (c) is *hierarchical training*.

horizontal training and *hierarchical training* paradigms. Observing that the works in Eshratifar et al. [2018] and Ren et al. [2019] only consider two levels in the mobile-edge-cloud hierarchical architecture—the device and cloud levels in Eshratifar et al. [2018] and the mobile and edge levels in Ren et al. [2019]—we believe that better performance can be achieved by fully utilizing the communication and computation resources of all three levels. In particular, as communication latency between mobile and edge levels is generally low and the computation resource at the cloud level is often abundant, a holistic framework that fully exploits the communication and computation resources of all three levels can definitely unleash the great potentials of mobile-edge-cloud computing for accelerating edge AI learning.

Motivated by the above observations, we propose a hierarchical training framework, abbreviated as HierTrain, which efficiently deploys the DNN training tasks over the mobile-edge-cloud levels and achieves minimum training time for fast edge AI learning. Specifically,

1. We develop a novel *hybrid parallelism* method, which is the key to HierTrain, to adaptively assign the DNN model layers and the data samples to the three levels by taking into account the communication and computation resource heterogeneity therein.

2. We formulate the problem of scheduling the DNN training tasks at both layer-granularity and sample-granularity. Solving this minimization problem enables us to reduce the training time.

3. We implement and deploy a hardware prototype over an edge device, an edge server, and a cloud server, and extensive experimental results demonstrate that HierTrain achieves superior performance.

We emphasize that, different from the federated learning schemes that focus on privacy-preserving edge AI training, in this chapter we promote HireTrain for addressing the important issue of edge AI training acceleration. This is due to the emerging demand that many edge AI applications (e.g., smart robots and industrial IoT) require both real-time performance and continuous learning capability of fast model updating with fresh sensing/input data samples and being adaptive to complex dynamic application environments. On the other hand, HierTrain is along the emerging line of promoting in-network model training such as edge learning for intelligent B5G networking [Murshed et al., 2019] for mitigating the significant overhead and latency of transferring the data of massive size to the cloud for remote model training.

5.2   BACKGROUND AND MOTIVATION

In general, there are three computing workers/nodes for DNN training in the mobile-edge-cloud hierarchical system: edge device, edge server, and cloud center, which have diverse communication and computation capacities. To jointly train a DNN, we need to determine how to split the training data samples and the trained DNN across the three workers. Below, we introduce two traditional methods, *model parallelism* and *data parallelism*, as well as our proposed *hybrid parallelism* method. The three parallelism methods are illustrated in Figure 5.2.

(1) Model Parallelism: Because a DNN is typically stacked by a sequence of distinct layers, it is natural to assign the layers to the workers, as shown in Figure 5.2a. In the *model parallelism* method, each worker holds multiple layers and is in charge of updating the corresponding model parameters. Therefore, when training the DNN with the back-propagation rule in the SGD algorithm [Bottou, 2010], the workers need to communicate to exchange the intermediate results. The works of JointDNN [Eshratifar et al., 2018] and JALAD [Li et al., 2018c] demonstrate the effectiveness of the *model parallelism* method. However, since the layers of the DNN are trained sequentially, when one worker is computing the others must stay idle. Thus, the computation resources are not fully utilized in the *model parallelism* method.

(2) Data Parallelism: The *data parallelism* method splits the data samples to the workers, trains one local copy of DNN in every worker, and forces the local DNNs to reach a consensus along the optimization process. To implement SGD, the workers need to exchange either the local stochastic gradients or the local model parameters from time to time, as depicted in Figure 5.2b. The works of Goyal et al. [2017] and You et al. [2017] show that the *data parallelism* method is able to accelerate the DNN training when the data are collected and split to multiple computing units within the cloud center. Nevertheless, the requirement of transmitting the local stochastic gradients or the local model parameters, whose dimensions are the same, leads to

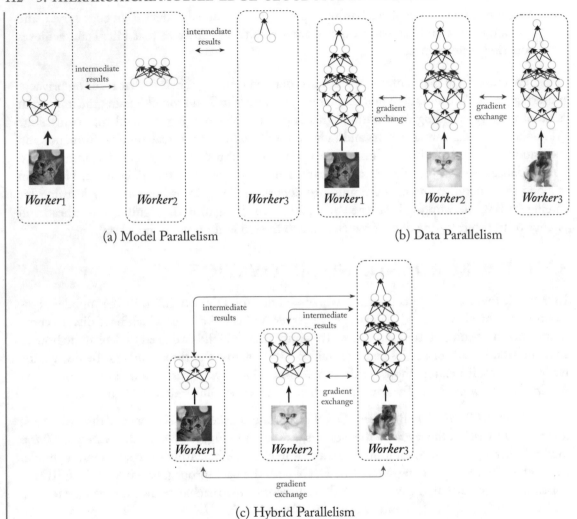

Figure 5.2: Illustration of the three parallelism methods. Each row of circles represents a layer in the trained DNN.

heavy communication overhead when the size of DNN is large. Therefore, the *data parallelism* method is not communication-efficient in the mobile-edge-cloud architecture.

(3) Hybrid Parallelism: We observe that the backend layers in most DNNs, such as CNNs, are fully connected layers and contain the majority of parameters. This fact motivates us to improve the *model parallelism* method through letting all the backend layers be trained by one worker while the frontend layers be trained by multiple workers. Therefore, the workers just

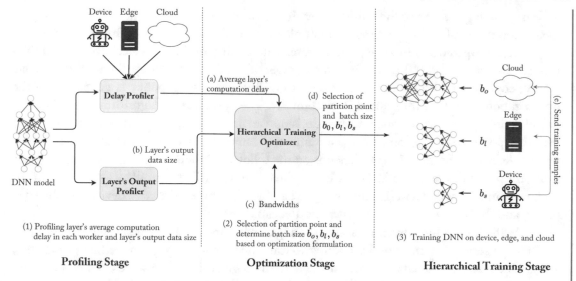

Figure 5.3: System overview of HierTrain.

need to exchange a small fraction of the local stochastic gradients or the local model parameters to train the frontend layers, as well as transmit the intermediate results to train the backend layers, thus the communication latency between workers is greatly reduced. As shown in Figure 5.2c, the backend layers are only trained by *worker₃*. Some frontend layers are trained by *worker₂* and *worker₃*, while some are jointly trained by all the workers. Meanwhile, similar to the data parallelism method, training data samples are split and assigned to all the workers according to their computing resource heterogeneity, to further balance the workloads across the device, edge and cloud.

In order to apply the *hybrid parallelism* method to accelerate the training of DNNs over the mobile-edge-cloud architecture, we need to optimize the assignments of the DNN layers and the data samples to the three workers. To this end, we propose HierTrain, a hierarchical training framework, as follows.

5.3   HIERTRAIN FRAMEWORK

In this section, we present the HierTrain framework, which jointly selects the best partition points of the given DNN model and determines the appropriate number of data samples delegated to different workers in a mobile-edge-cloud hierarchy. Figure 5.3 presents the system overview of the HierTrain framework, which consists of three stages: profiling, optimization, and hierarchical training.

At the **profiling stage**, HierTrain performs two initialization steps: (i) profiling the average execution time of different model layers in the device, edge, and cloud workers, respectively;

and (ii) profiling the size of the output for each layer in the model. We conduct the profiling by measuring these values in run-time for multiple times, and calculating their mean values.

At the **optimization stage**, the hierarchical training optimizer selects the best DNN model partition points and determines the number of training samples for the workers of edge device, edge server and cloud center, respectively. This scheduling policy is generated by the optimization algorithm introduced in Section 5.4. The optimization algorithm minimizes the DNN training time with respect to five decision variables m_s, m_l, b_o, b_s, b_l (m_s, m_l represent partition points, b_o, b_s, b_l represent the number of samples processed on each worker, which will be defined in Section 5.4). It depends on the following inputs: (i) the profiled average execution time of different model layers in the three workers; (ii) the profiled size of output for each layer in the model; and (iii) the available bandwidth between the edge device and the edge server, and that between the edge server and the cloud center.

At the **hierarchical training stage**, the edge device first sends the delegated data samples to the edge server and the cloud according to the scheduling policy given in the optimization stage. Once the needed data samples are at hand, the edge device, edge server, and cloud center start their scheduled training tasks (i.e., the assigned model training modules) immediately, and perform collaborative model training in a hierarchical manner.

Note that the hierarchical training stage depicted in Figure 5.3 only shows one possible scheduling policy, in which the cloud trains the full model while the edge server and the edge device only train parts of the model. This scheduling policy is suitable for the scenario that the bandwidth between edge device and cloud center is in good condition. However, when the network bandwidth becomes the bottleneck, the scheduling policy may choose the edge server or the edge device to train the full model. In the next section, we will elaborate on how the data samples and model layers are partitioned.

5.4 PROBLEM STATEMENT OF POLICY SCHEDULING

5.4.1 TRAINING TASKS IN HIERTRAIN

We consider that a DNN is stacked by a sequence of distinct layers, and the output of one layer feeds into the input of the next layer. Our goal is to reduce the overall training time in the mobile-edge-cloud environment. Toward this end, we first define three types of training tasks, depicted in Figure 5.4 and explained as follows.

TASK O (Original Task): Training the full DNN with b_o data samples.
TASK S (Short Task): Training m_s consecutive layers from layer 1 to layer m_s with b_s data samples.
TASK L (Long Task): Training m_l consecutive layers from layer 1 to layer m_l with b_l data samples.
Here m_s and m_l are positive integers, and we assume $m_s \leq m_l \leq N$ (N is the total number of layers in the DNN model).

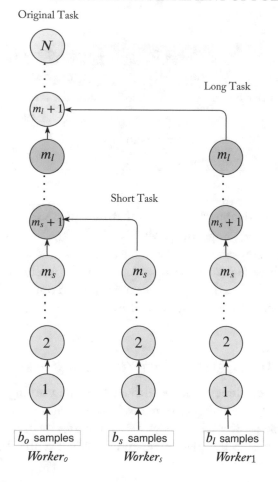

Figure 5.4: $worker_o$, $worker_s$, $worker_l$ use b_o, b_s, b_l data samples as inputs, respectively. Layer 1 to layer m_s are executed in parallel over the three workers, layer $m_s + 1$ to layer m_l are executed in parallel over $worker_o$ and $worker_l$, and the rest of layers are executed on $worker_o$.

The key motivations of defining the three task types above are as follows. On one hand, only TASK O contains the most backend layers (e.g., fully connected layers in many DNNs) that typically have the majority of the parameters, and this helps to reduce the communication overheads for parameter exchange across different tasks. On the other hand, TASK O, L, and S all contain the frontend layers (e.g., convolution layers in many DNNs) that are often computationally intensive, and this also helps to exploit the computing resources of different workers in parallel to accelerate the DNN training. Furthermore, we have the flexibility to optimize the computing workloads of different tasks by varying their input data sample sizes.

Table 5.1: List of notations

Parameter	Description
$L_{j,i}^f$	Forward time to handle 1 sample for layer i on worker j
$L_{j,i}^b$	Backward time to handle 1 sample for layer i on worker j
$L_{j,i}^u$	Weight update time for layer i on worker j
MP_i	Number of parameters in layer i
MO_i	Output size of layer i in forward phase

In the following, we denote the workers that execute TASK O, TASK S, and TASK L as $worker_o$, $worker_s$, and $worker_l$, respectively. We also denote the profiling values $L_{j,i}^f$, $L_{j,i}^b$, $L_{j,i}^u$, and MP_i, MO_i. Their meanings are shown in Table 5.1.

By defining the three task structures, we have rich flexibility in optimizing the training workloads across the edge device, the edge server and the cloud center by tuning the sizes of their data samples and assigned model layers, tailored to their computation resources and network conditions.

5.4.2 TRAINING PROCEDURE IN HIERTRAIN

Based on the above-defined three tasks, we elaborate on the training procedure in HierTrain as follows. First, the scheduling policy determines how to assign the model layers and the data samples to the three workers. Second, the edge device initiates the training procedure and sends the partitioned data samples to the edge server and the cloud center. Last, the following three phases are executed iteratively.

(1) Forward: worker_s executes the forward phase (i.e., inference through the DNN model to obtain the current model loss) over the assigned layers, using a mini-batch b_s of data samples. Once completing the forward phase over the assigned layers, $worker_s$ sends the output to $worker_o$. Then, $worker_o$ proceeds to execute the forward phase over the rest of layers. $worker_l$ acts the same as $worker_s$, using a mini-batch b_l of data samples. $worker_o$ also executes the forward phase, but over all the layers and using a mini-batch b_o of data samples. When the forward phase ends, $worker_o$ collects the model losses from $B = b_s + b_l + b_o$ data samples.

(2) Backward: For each data sample, $worker_o$ starts the backward phase (i.e., back-propagation using the loss to obtain the stochastic gradients) from the last layer of the DNN. If the data sample belongs to $worker_o$, then $worker_o$ executes the full backward phase. If the data sample belongs to $worker_l$, then $worker_o$ sends the intermediate results to $worker_l$ upon reaching layer $m_l + 1$, and $worker_l$ proceeds to execute the backward phase over the rest of layers. The same rule applies to $worker_s$, except that $worker_o$ sends the intermediate results to $worker_s$

upon reaching layer $m_s + 1$. When the backward phase ends, every worker obtains the stochastic gradients of the assigned layers.

(3) Weight Update: worker$_l$ and *worker$_s$* send the computed stochastic gradients to *worker$_o$*. Then *worker$_o$* averages the stochastic gradients layer-wise, and sends the averaged stochastic gradients to *worker$_l$* and *worker$_s$* according to the layers assigned to them. With these stochastic gradients, the three workers update the weights of their assigned layers independently.

5.4.3   MINIMIZATION OF TRAINING TIME

The core of HierTrain is a scheduling policy that determines how the model layers and the data samples are assigned to the three workers. The primary goal is to minimize the training time, which is determined by the computation and communication latencies. To analyze these two quantities, we assume that the DNN has N layers and the size of each data sample is Q bits.

(1) Computation Latency: Recall that the DNN training procedure is divided into three phases: forward, backward, and weight update. In the forward and backward phases, the amount of computation is proportional to the number of processed data samples [Devarakonda et al., 2017]. We denote $T_{j,i,b,forward}$ and $T_{j,i,b,backward}$ as the computation latencies of executing layer i on *worker$_j$* with b input data samples in the forward and backward phases, respectively. Here $j \in \{o, s, l\}$, $i \in \{1, 2, \ldots, N\}$, and $b \in \{b_o, b_o + b_l, b_o + b_l + b_s\}$. Then we have

$$T_{j,i,b,forward} = bL_{j,i}^{f}, \tag{5.1}$$
$$T_{j,i,b,backward} = bL_{j,i}^{b}. \tag{5.2}$$

The computation latency $T_{j,update}$ of the weight update phase on a worker $j \in \{o, s, l\}$ is the summation of the execution time over the involved layers, given by

$$T_{j,update} = \sum_{i=1}^{m_j} L_{j,i}^{u}. \tag{5.3}$$

(2) Communication Latency: The workers are bidirectionally connected with each other. For example, the edge device and the edge server are connected with the high-speed wireless local-area-network (WLAN) link, while the edge server and the cloud center are connected with the bandwidth-limited wide-area-network (WAN) link. Let $B_{o,s}$ denote the bandwidth between *worker$_o$* and *worker$_s$*, $B_{o,l}$ the bandwidth between *worker$_o$* and *worker$_l$*, $B_{s,l}$ the bandwidth between *worker$_s$* and *worker$_l$*. The communication latency is the ratio of the transferred data size and the bandwidth between two workers, as

$$T_{communication} = \frac{DataSize}{Bandwidth}. \tag{5.4}$$

(3) Training Time: As depicted in Figure 5.4, *worker$_o$*, *worker$_s$*, *worker$_l$*, use b_o, b_s, b_l data samples as inputs, respectively. Layer 1 to layer m_s are executed in parallel over the three

workers, layer $m_s + 1$ to layer m_l are executed in parallel over $worker_o$ and $worker_l$, and the rest of layers are executed on $worker_o$. Below we calculate the training time, beginning with those in the forward and backward phases.

Let $T_{forward}^1$ and $T_{backward}^1$ denote the latencies of executing layers between 1 and m_s over the three workers in the forward and backward phases, respectively, given by

$$T_{forward}^1 = \max\{T_{o,input} + \sum_{i=1}^{m_s} T_{o,i,b_o,forward}, T_{s,input} + \sum_{i=1}^{m_s} T_{s,i,b_s,forward} + T_{s,output},$$

$$T_{l,input} + \sum_{i=1}^{m_s} T_{l,i,b_l,forward}\}, \tag{5.5}$$

$$T_{backward}^1 = \max\{\sum_{i=1}^{m_s} T_{o,i,b_o,backward}, \sum_{i=1}^{m_s} T_{s,i,b_s,backward} + T_{s,grad}, \sum_{i=1}^{m_s} T_{l,i,b_l,backward}\}, \tag{5.6}$$

where $T_{j,input}$ is the communication latency of $worker_j$ to receive b_j data samples, $j \in \{o, s, l\}$. We use (5.4) to calculate $T_{j,input}$, where $DataSize = b_j \times Q$ and $Bandwidth$ is the bandwidth between the edge device and $worker_j$. $T_{s,output}$ represents the communication latency of $worker_s$ to transmit its forward output to $worker_o$. Recall that MO_{m_s} is the output size of layer m_s in the forward phase for one data sample, b_s is the number of data samples of $worker_s$, and $B_{o,s}$ is the bandwidth between $worker_o$ and $worker_s$. Then according to (5.4), $T_{s,output} = \frac{b_s \times MO_{m_s}}{B_{o,s}}$. $T_{s,grad}$ represents the communication latency of $worker_o$ to send the intermediate results to $worker_s$ in the backward phase. The size of the intermediate results is equal to the output data of layer m_s in forward phase. Thus, $T_{s,grad} = T_{s,output}$.

Denote $T_{forward}^2$ and $T_{backward}^2$ as the latencies of executing layers between $m_s + 1$ and m_l over $worker_o$ and $worker_l$ in the forward and backward phases, respectively, given by

$$T_{forward}^2 = \max\{\sum_{i=m_s+1}^{m_l} T_{o,i,b_o+b_s,forward}, \sum_{i=m_s+1}^{m_l} T_{l,i,b_l,forward} + T_{l,output}\}, \tag{5.7}$$

$$T_{backward}^2 = \max\{\sum_{i=m_s+1}^{m_l} T_{o,i,b_o+b_s,backward}, \sum_{i=m_s+1}^{m_l} T_{l,i,b_l,backward} + T_{l,grad}\}, \tag{5.8}$$

where $T_{l,output}$ is the communication latency of $worker_l$ to transmit its forward output to $worker_o$, given by $T_{l,output} = \frac{b_l \times MO_{m_l}}{B_{o,l}}$. $T_{l,grad}$ represents the communication latency of $worker_o$ to send the intermediate results to $worker_l$ in the backward phase, it is equal to $T_{l,output}$.

Denote $T^3_{forward}$ and $T^3_{backward}$ as the latencies of executing layers between $m_l + 1$ and N over $worker_o$ in the forward and backward phases, respectively, given by

$$T^3_{forward} = \sum_{i=m_l+1}^{N} T_{o,i,b_o+b_s+b_l,forward}, \tag{5.9}$$

$$T^3_{backward} = \sum_{i=m_l+1}^{N} T_{o,i,b_o+b_s+b_l,backward}. \tag{5.10}$$

Now we consider the training time in the weight update phase. After the backward phase finishes, $worker_l$ and $worker_s$ send the stochastic gradients to $worker_o$. Then $worker_o$ sends the averaged stochastic gradients to $worker_l$ and $worker_s$ according to the layers assigned to them, and the three workers update the weights of their assigned layers. The total time cost in the weight update phase is denoted as T_{update}, given by

$$T_{update} = \max\{T_{o,update}, T_{s,update}, T_{l,update}\} + \max\{T_{s,weightgrad}, T_{l,weightgrad}\}, \tag{5.11}$$

where $T_{j,update}$ is the computation latency of the weight update phase on $worker_j$, $j \in \{o, s, l\}$, as defined in (5.3). $T_{s,weightgrad}$ and $T_{l,weightgrad}$ represent the communication latencies of $worker_s$ and $worker_l$ to send the stochastic gradients to and receive the updated weights from $worker_o$, respectively. For layer i, the sizes of the stochastic gradients and the updated weights are both MP_i. Therefore, we have $T_{s,weightgrad} = \frac{2}{B_{o,s}} \sum_{i=1}^{m_s} MP_i$ and $T_{l,weightgrad} = \frac{2}{B_{o,l}} \sum_{i=1}^{m_l} MP_i$.

(4) Minimization of Training Time: Therefore, the time of training the DNN for one iteration, including both computation and computation, is given by

$$T_{total} = \sum_{k=1}^{3} (T^k_{forward} + T^k_{backward}) + T_{update}, \tag{5.12}$$

in which the number of used data samples is

$$B = b_o + b_s + b_l. \tag{5.13}$$

Here B is the predefined batch size, while b_o, b_s, and b_l are decision variables.

The number of layers m_s and m_l for TASK S and TASK L are also decision variables. It is possible in some scenarios that m_s or m_l can equal to 0, meaning that $worker_s$ or $worker_l$ will not participate in the DNN training procedure. For these scenarios, we do not assign any data samples to $worker_s$ or $worker_l$, such that $b_s = 0$ or $b_l = 0$. To characterize these connections, we introduce constraints

$$0 \le b_s \le m_s B, \tag{5.14}$$
$$0 \le b_l \le m_l B. \tag{5.15}$$

Algorithm 5.5 HierTrain Algorithm

1: **Input:**

    1. $L_{k,i}^f, L_{k,i}^b, L_{k,i}^u, k \in \{d, e, c\}$: profiling values of device, edge, cloud

    2. BW_{de}, BW_{ec}: bandwidth of device-edge and edge-cloud

    3. MP_i: layer i parameters data size

    4. MO_i: layer i output data size

    Output: optimal solution $\{m_s^*, m_l^*, b_o^*, b_s^*, b_l^*\}$

2: **Initialization:** $T_{total,minimum} = MAX$ ▷ *MAX* is an infinite number

3: **for** $\{worker_o, worker_s, worker_l\} \leftarrow permutation\{d, e, c\}$ **do** ▷ map {device, edge, cloud} to $\{node_o, node_s, node_l\}$

4:      **for** $m_s = 0 \to N$ **do**

5:          **for** $m_l = m_s \to N$ **do**

6:              Solve problem $\mathcal{P}_1$ to get $\{b_o, b_s, b_l\}$

7:              $\{b_o, b_s, b_l\} \leftarrow Round(b_o, b_s, b_l)$ ▷ rounding b_o, b_s, b_l to integers

8:              Calculate T_{total} according to (5.12)

9:              **if** $T_{total} < T_{total,minimum}$ **then**

10:                  $\{m_s^*, m_l^*, b_o^*, b_s^*, b_l^*\} = \{m_s, m_l, b_o, b_s, b_l\}$

11:                  $T_{total,minimum} = T_{total}$

Return: $\{m_s^*, m_l^*, b_o^*, b_s^*, b_l^*\}$

When $m_s = 0$ or $m_l = 0$, (5.14) or (5.15) ensures that $b_s = 0$ or $b_l = 0$. Otherwise, if m_s or m_l is any positive integer, (5.14) or (5.15) automatically satisfies due to (5.13).

In summary, when $worker_s$, $worker_l$, and $worker_o$ have been fixed, to minimize the training time, HierTrain solves the following optimization problem:

$$\mathcal{P}_1 : \quad \underset{\{b_o, b_s, b_l, m_s, m_l\}}{minimize} \quad T_{total}$$
$$s.t. \quad b_o + b_s + b_l = B,$$
$$0 \le b_s \le m_s B,$$
$$0 \le b_l \le m_l B,$$

where the decision variables b_o, b_s, b_l, m_s, m_l are all nonnegative integers. Since there are six possible mappings between $worker_s$, $worker_l$, $worker_o$ and the edge device, the edge server, the cloud center, we can enumerate all the mappings, calculate the optimal scheduling policy $\{b_o, b_s, b_l, m_s, m_l\}$ for each mapping, and then find the global optimal scheduling policy. The next section gives details of the proposed algorithm.

5.5   OPTIMIZATION OF POLICY SCHEDULING

Note that even when $worker_s$, $worker_l$, and $worker_o$ have been fixed, solving $\mathcal{P}_1$ is still challenging because: (i) in the objective T_{total}, the terms of T_{update}, $T_{forward}^k$, and $T_{backward}^k$, where $k = 1, 2, 3$, all contain summations with the numbers of summands determined by m_s and m_l; and (ii) the decision variables b_o, b_s, b_l, m_s, m_l are all integers.

To address the first challenge, we observe that when m_s and m_l are fixed, $\mathcal{P}_1$ will become a standard integer linear programming (ILP) problem and is relatively easier to solve. Motivated by this observation, we enumerate the values of m_s and m_l, solve the resulting ILP problems, and then find the best one among the ILP solutions. This enumeration is feasible because the numbers of layers m_s and m_l are often modest in practice (such as AlexNet: 8 layers, VGG-16: 16 layers, GoogleNet: 22 layers, MobileNet: 28 layers).

To address the second challenge, for each ILP problem, we relax the integer variables to real ones, solve the relaxed linear programming (LP) problem, and then round the solution to integers. To be specific, the relaxed LP problem can be efficiently solved with CPLEX, Gurobi, or CVXPY. Although these optimization solvers can solve ILP problem directly, we choose to convert the ILP problem to LP problem the reason is that these solvers solve LP problem are much faster than solve ILP problem. Further, the **rounding operation** works as follows. Given a real solution (b_o, b_s, b_l) of the relaxed LP problem, we divide them into integer parts $int(b_j)$ and fraction parts $frac(b_j)$, $j \in \{o, s, l\}$, and then sort the fraction parts in a descending order. For h_j with the largest fraction part, we let $b_j^* = int(b_j) + 1$, while for the other two b_j, $b_j^* = int(b_j)$. If $b_o^* + b_s^* + b_l^* = B$ is satisfied, then the rounding operation ends. Otherwise, for the two b_j with the largest fraction parts, we let $b_j^* = int(b_j) + 1$, while for the other b_j, $b_j^* = int(b_j)$. The constraint $b_o + b_s + b_l = B$ can be satisfied after at most two steps.

So far, we have solved $\mathcal{P}_1$ given that $worker_s$, $worker_l$, and $worker_o$ have been fixed. In order to deploy the DNN training task over the device-edge-cloud environment, we still need to find the best mapping strategy between the device, edge and cloud workers and $worker_o$, $worker_s$, and $worker_l$. As illustrated in Figure 5.5, since the overall number of mappings is only 6, we can enumerate all the mapping, find a candidate optimal scheduling policy for each mapping, and then choose the best mapping strategy with the minimum training time. The algorithm is outlined in **Algorithm** 5.5.

As shown in the Table 5.2, in order to verify the efficiency of our algorithm, we list the algorithm running time based on some common deep networks configuration. All results are obtained on a desktop computer equipped with an Intel Core(TM) i7-6700 3.4 GHz with 8 GB RAM running Linux. We use python as programming language and CPLEX as optimization problem solver. From Table 5.2, we see that the proposed algorithm runs very fast, and in practice its running time can be ignored compared with the long DNN training time.

Table 5.2: Algorithm running time

LeNet	AlexNet	VGG-16	VGG-19	googLeNet	ResNet-34
0.52 s	1.48 s	3 s	4 s	5.3 s	12 s

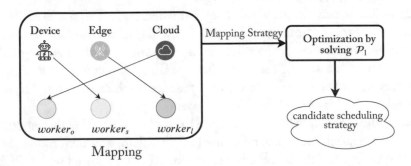

Figure 5.5: Each mapping strategy corresponds to a candidate optimal scheduling policy.

5.6   PERFORMANCE EVALUATION

5.6.1   DATASET AND MODELS

We evaluate HierTrain by training two well-known CNNs for image classification tasks. The first CNN is LeNet-5 [LeCun et al., 1998], and we train it with the CIFAR-10 dataset [Krizhevsky et al., 2009]. CIFAR-10 contains 50,000 training images and 10,000 testing images, each of which has 10 labels. The second CNN is AlexNet [Krizhevsky et al., 2012], which is more complicated than LeNet-5. We train AlexNet on the tiny ImageNet dataset. The tiny ImageNet dataset has 200 classes, while each class has 500 training images, 50 validation images, and 50 testing images.

5.6.2   EXPERIMENTAL SETUP

We use a Raspberry Pi 3 tiny computer to act as an edge device. The Raspberry Pi 3 has a quad-core ARM processor at 1.2 GHz with 1 GB of RAM. We use an Intel NUC, a small but powerful mini PC which is equipped with a four Intel Cores (TM) i3-7100U with 8 GB of RAM, to emulate the edge server. Unless specifically indicated, we only use one core of the edge server in our experiments. This is to simulate the application scenarios where the edge server has to serve multiple edge devices and each edge device cannot occupy all the computation resource of the edge server. The cloud center is a Dell Precision T5820 Tower workstation with 16 Intel Xeon processor at 3.7 GHz and with 30 GB of RAM, and equipped with NVIDIA GPU GeForce GTX 1080 Ti. The computation capability of the cloud center is one order magnitude higher than those of the edge device and the edge server. All the three workers run the Ubuntu system, and we use Linux Traffic Control on them to emulate constrained network bandwidths.

There are many existing open-source platforms for training CNNs, such as Tensor-Flow [Abadi et al., 2016], Theano [Bergstra et al., 2010], MXNet [Chen et al., 2015a], Py-Torch [Paszke et al., 2017], and Chainer [Tokui et al., 2015]. Among them we choose Chainer because it is flexible and able to leverage dynamic computation graphs, which facilitates the application of the proposed *hybrid parallelism* method.

5.6.3   BASELINES

To elucidate the performance of the proposed HierTrain framework, we consider the following baselines in the experimental evaluation.

(1) All-Edge: The edge device transmits all the training data samples to the edge server, and the edge server completes the DNN training.

(2) All-Cloud: The edge device transmits all the training data samples to the cloud center, and the cloud center completes the DNN training.

(3) JointDNN [Eshratifar et al., 2018]: The edge device and the cloud center jointly train the DNNs.

(4) JointDNN+: We extend JointDNN to train the DNNs in the mobile-edge-cloud architecture. Following the design of JointDNN, the scheduling in JointDNN+ is by solving a shortest path problem over a graphic model.

(5) JALAD [Li et al., 2018c]: The edge server and the cloud center jointly train the DNNs. A data compression strategy is applied to reduce the edge-cloud transmission latency. In our experiments we set the number of bits c used in data compression as 8.

5.6.4   RESULTS

We now elaborate the evaluation results from the following different perspectives.

(1) Model Validity: We first validate the formulated model that captures the execution delay of one iteration in training a DNN. Using the same scheduling policy, we obtain the real latency measured from the experiment and the theoretical latency, both in training AlexNet. As is shown in Figure 5.6, the real and theoretical latencies highly match.

(2) Comparison with All-Edge and All-Cloud: Next we compare HierTrain with the two baselines, All-Edge and All-Cloud, by fixing the mobile-edge bandwidth to 5 Mbps and varying the edge-cloud bandwidth from 1.5–5 Mbps. Figure 5.7 shows the average per-iteration time to train AlexNet. The time cost of All-Cloud decreases as the edge-cloud bandwidth increases, while that of All-Edge remains unchanged. HierTrain outperforms, and achieves up to 2.3×

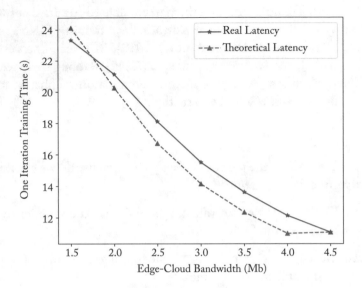

Figure 5.6: Comparison of real and theoretical latencies of training AlexNet.

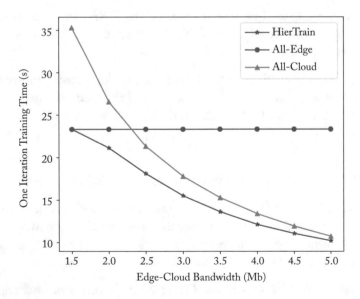

Figure 5.7: Per-iteration training time of AlexNet for HierTrain, All-Edge, and All-Cloud under different bandwidths.

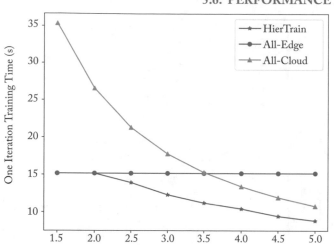

Figure 5.8: Per-iteration training time of LeNet-5 for HierTrain, All-Edge, and All-Cloud under different bandwidths.

and 4.5× speedup comparing to All-Edge and All-Cloud, respectively. Similar observations can be found in training LeNet-5, as depicted in Figure 5.8. HierTrain is the best among the three schemes, achieves up to 1.7× and 6.9× speedup comparing to All-Edge and All-Cloud, respectively.

(3) Comparison with JointDNN, JointDNN+, and JALAD: Now we conduct experiments to compared HierTrain with the three baselines: two state-of-the-art methods JointDNN and JALAD, as well as JointDNN+ that extends JointDNN to the mobile-edge-cloud architecture. The results on training AlexNet and LeNet-5 are demonstrated in Figures 5.9 and 5.10, respectively. Observe that HierTrain outperforms both JointDNN and JointDNN+. Among these two baselines, JointDNN+ is better than JointDNN because it can utilize the edge server when the edge-cloud bandwidth is as low as 1.5 Mbps or 2 Mbps. When the edge-cloud bandwidth becomes larger, both JointDNN and JointDNN+ choose to run the training tasks in the cloud center.

Figure 5.9 also compares HierTrain and JALAD in training AlexNet. When the edge-cloud bandwidth ranges from 1.5–2 Mbps, JALAD performs better than HierTrain. The reason is that the data compression strategy of JALAD can largely reduce the amount of transmitted data between the edge server and the cloud center. This makes JALAD advantageous in the low bandwidth condition as the communication time cost is the dominating factor in the overall delay. However, when the bandwidth increases, the benefit from reducing communication delay with data compression degrades sharply, and HierTrain outperforms JALAD. In Figure 5.10 that shows the experimental results of training LeNet-5, the curves of JALAD and JointDNN+

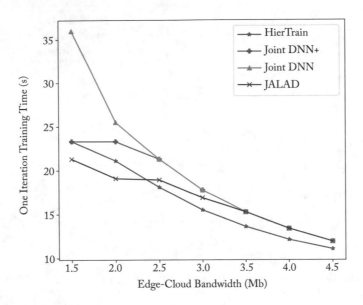

Figure 5.9: Per-iteration training time of AlexNet for HierTrain, JointDNN, JointDNN+, and JALAD under different bandwidths.

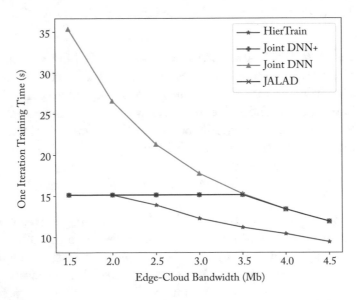

Figure 5.10: Per-iteration training time of LeNet-5 for HierTrain, JointDNN, JointDNN+, and JALAD under different bandwidths.

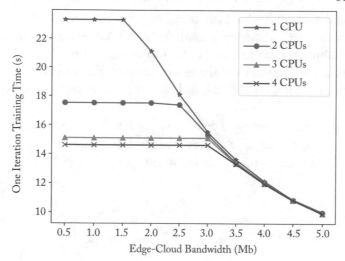

Figure 5.11: Effect of varying computation capability of edge server in HierTrain under different bandwidths.

overlap, because their scheduling policies are the same in the scenario—they are the same as the All-Edge strategy in the low bandwidth condition and the All-Cloud strategy in the high bandwidth condition.

(4) Effect of varying edge server resources: Finally, we investigate the performance of HierTrain when the computation capability at the edge server changes. We consider training AlexNet, while keep the mobile-edge bandwidth as 5 Mbps and the edge-cloud bandwidth as 3.5 Mbps. We use docker to control the CPU cores used in the training process. As shown in Figure 5.11, when the edge-cloud bandwidth is very low (≤ 1.5 Mbps), improving the computation capability of the edge server can speedup the training process. This performance gain shrinks when the computation capability of the edge server keeps increasing. To be specific, varying from 1–2 CPUs leads to large speedup, while varying from 3–4 CPUs yields insignificant speedup. When the edge-cloud bandwidth is sufficiently large (≥ 3 Mbps), the computation capability of the edge server does not influence the overall performance. The reason for this phenomenon is that when the edge-cloud bandwidth is sufficiently large, the optimal policy is training on the cloud.

5.7   SUMMARY

In this chapter, we study the problem of accelerating the training procedure of DNNs on the device-edge-cloud architecture. To this end, first, we present a novel *hybrid parallelism* method for training DNNs. Second, in order get scheduling policy of using *hybrid parallelism* method to train DNNs on the device-edge-cloud environment, we formulate the problem of computation scheduling of training DNNs at layer-granularity and sample-granularity as a minimization optimization programming problem, and solve it to get the scheduling policy. In addition, we test HierTrain in the real hardware and the results show that it could obviously outperform the naive policy such as all-edge and all-cloud, and also outperform exist prior work like JointDNN and JALAD.

For future work, we are going to generalize the HierTrain framework to the application scenarios in multi-device and multi-edge environments, in which the federated learning across multi-devices and the device-to-edge association are interesting and challenging.

CHAPTER 6

Edge Intelligence via Model Inference

Chapters 2–5 focus on efficient training of EI models. Needless to say, real-time inference at the edge can be equally critical for enabling high-quality edge intelligence service deployment. In this chapter, we discuss the DNN model inference at the edge, including the architectures, key performance indicators, enabling techniques, and existing systems and frameworks.

6.1 ARCHITECTURES

Besides the common cloud-based and device-cloud inference architectures, we further define several major edge-centric inference architectures and classify them into four modes, namely edge-based, device-based, edge-device, and edge-cloud, as illustrated in Figure 6.1.

Specifically, we demonstrate four different DNN model inference modes in Figure 6.1, and describe the main workflow of each mode as follows. It is worth noting that the four edge-centric inference modes can be adopted in a system simultaneously to carry out complex AI model inference tasks (e.g., Cloud-Edge-Device hierarchy), by efficiently pooling heterogeneous resources across a multitude of end devices, edge nodes, and cloud.

6.1.1 EDGE-BASED

In Figure 6.1a, Device A is in the edge-based mode, where the device receives the input data and then sends it to the edge server. After the DNN model inference is finished at the edge server, the prediction results will be returned to the device. Under this inference mode, since the DNN model is on the edge server, it is easy to implement the application on different mobile platforms. Nevertheless, one main disadvantage is that the inference performance depends on network bandwidth between the device and the edge server.

6.1.2 DEVICE-BASED

In Figure 6.1b, Device B is in the device-based mode, where the mobile device obtains the DNN model from the edge server and performs the model inference locally. During the inference process, the mobile device does not communicate with the edge server. To achieve reliable inference, it requires a large amount of resources such as CPU, GPU, and RAM on the mobile device. The performance depends on the local device itself.

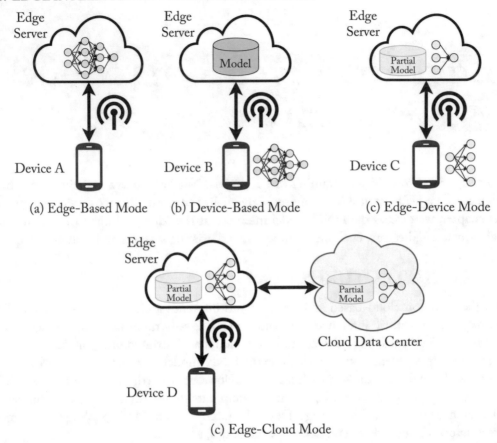

Figure 6.1: Major edge-centric inference modes: edge-based, device-based, edge-device, and edge-cloud.

6.1.3   EDGE-DEVICE

In Figure 6.1c, Device C is in the edge-device mode, where the device first partitions the DNN model into multiple parts according to the current system environmental factors such as network bandwidth, device resource, and edge server workload, and then executes the DNN model up to a specific layer with the intermediate data sent to the edge server. The edge server will next execute the remain layers and send the prediction results back to the device. Compared to the edge-based mode and the device-based mode, the edge-device mode is more reliable and flexible. It may also require a significant amount of resources on the mobile device because the convolution layers at the front position of a DNN model are generally computational-intensive.

6.1.4 EDGE-CLOUD

In Figure 6.1d, Device D is in the edge-cloud mode, which is similar to the edge-device mode and suitable for the case where the device is highly resource-constrained. In this mode, the device is responsible for input data collection and the DNN model is executed through edge-cloud synergy. The performance of this model heavily depends on the network connection quality.

6.2 KEY PERFORMANCE INDICATORS

To evaluate the service quality of model inference for edge intelligence, we introduce the following five metrics.

6.2.1 LATENCY

Latency refers to the time spent in the whole inference process, including pre-processing, model inference, data transmission, and post-processing. For some real-time intelligent mobile applications (e.g., AR/VR mobile gaming and intelligent robots), they usually have stringent deadline requirements such as 1 ms latency. Latency indicator is affected by many factors, including the resources on edge devices, the way of data transmission, and the way to execute the DNN model.

6.2.2 ACCURACY

Accuracy refers to the ratio of the number of the input samples that get the correct prediction from inference to the total number of input samples, reflecting the performance of the DNN models. For some mobile applications requiring a high level of reliability, such as self-driving car and face authentication, the ultra-high accuracy on the DNN model inference is essential. Besides the DNN model's own inference capability, the inference accuracy depends on the speed of feeding the input data to the DNN model. For a video analytics application, under a fast feeding rate, some input samples may be skipped due to the edge device's constraint resources, causing a drop in accuracy.

6.2.3 ENERGY

To execute a DNN model, compared with the edge server and the cloud data center, the end devices are usually battery-limited, while the computation and communication overheads of DNN model inference bring a large amount of energy consumption. For an edge intelligence application, energy efficiency is of great importance and affected by the size of DNN model and the resources on edge devices.

6.2.4 PRIVACY

The IoT and mobile devices generate a huge amount of data, which could be privacy sensitive. Thus, it is also important to protect privacy and data security near the data source for an edge intelligence application during the model inference stage. Privacy protection depends on the way of processing the original data.

6.2.5   COMMUNICATION OVERHEAD

Except for the device-based mode, the communication overhead affects the inference performance of the other modes greatly. It is necessary to minimize the overhead during the DNN model inference in an edge intelligence application, particularly the expensive wide-area network bandwidth usage for the cloud. Communication overhead here mainly depends on the mode of DNN inference and the available bandwidth.

6.2.6   MEMORY FOOTPRINT

Optimizing the memory footprint of performing DNN model inference on mobile devices is also crucial. On one hand, typically, a high-precision DNN model is accompanied by millions of parameters, which is very hungry for the hardware resources of mobile devices. On the other hand, unlike high-performance discrete GPUs on the cloud data center, there is no dedicated high-bandwidth memory for mobile GPUs on mobile devices [Wu et al., 2019]. Moreover, mobile CPUs and GPUs typically compete for shared and scarce memory bandwidth. For the optimization of the DNN inference at the edge side, memory footprint is a non-negligible indicator. Memory footprint is mainly affected by the size of the original DNN model and the way of loading the tremendous DNN parameters.

6.3   ENABLING TECHNOLOGIES

In this section, we review several enabling technologies to improve one or more of the aforementioned key performance indicators for edge intelligence model inference. Table 6.1 summarizes the highlights of each enabling technology.

6.3.1   MODEL COMPRESSION

To alleviate the tension between resource hungry DNNs and resource-poor end devices, DNN compression has been commonly adopted to reduce the model complexity and resource requirement, enabling local, on-device inference which in turn reduces the response latency and has fewer privacy concerns. That is to say, model compression method optimizes the above four indicators, i.e., latency, energy, privacy, and memory footprint. Various DNN compression techniques have been proposed, including weight pruning, data quantization, and compact architecture design.

Weight pruning is one of the most widely adopted techniques for model compression. This technique removes redundant weights (i.e., connections between neurons) from a trained DNN. Specifically, it first ranks the neurons in the DNN according to how much the neuron contributes, and then removes the low-ranking neurons to reduce the model size. Since removing

Table 6.1: Technologies for distributed DNN inference at the edge

Technology	Highlights	Related Work
Model Compression	• Weight pruning and quantization to reduce storage and computation	Chen et al. (2016); Han et al. (2015a,b,); Lane et al. (2016); Liu et al. (2018); Oh et al. (2018); Reagen et al. (2016); Yang et al. (2017)
Model Partition	• Computation offloading to the edge server or mobile devices • Latency- and energy-oriented optimization	Huang et al. (2019); Jeong et al. (2018b); Kang et al. (2017b); Ko et al. (2018); Li et al. (2018b,c); Mao et al. (2017a,b); Zeng et al. (2019); Zhao et al. (2018b)
Model Early-Exit	• Partial DNNs model inference • Accuracy-aware	Bolukbasi et al. (2017); Leroux et al. (2017); Li et al. (2018b,d); Lo et al. (2017); Teerapittayanon et al. (2016, 2017); Zeng et al. (2019)
Edge Caching	• Fast response toward reusing the previous results of the same task	Chen et al. (2015b); Drolia et al. (2017a,b); Guo et al. (2018); Venugopal et al. (2018)
Input Filtering	• Detecting difference between inputs, avoiding abundant computation	Canel et al., (2018); Jain et al. (2018); Kang et al. (2017a); Wang et al. (2018a); Zhang et al. (2018a)
Model Selection	• Input-oriented optimization • Accuracy-aware	Jiang et al. (2018b); Park et al. (2015); Shu et al. (2018); Stamoulis et al. (2018); Taylor et al. (2018)
Support for Multi-Tenancy	• Scheduling multiple DNN-based task • Resource-efficient	Fang et al. (2018a,b); Hung et al. (2018); Jiang et al. (2018a,b); Mathur et al. (2017); Narayanan et al. (2018)
Application-Specific Optimization	• Optimizations for the specific DNN-based application • Resource-efficient	Jiang et al. (2018b); Ran et al. (2018)

neurons damages the accuracy of the DNN, how to reduce the network size meanwhile preserving the accuracy is the key challenge. For modern large-scale DNNs, a pilot research [Han et al., 2015b] in 2015 tackled this challenge by applying a magnitude-based weight pruning method. The basic idea of this method is as follows: first remove small weights whose magnitudes are below a threshold (e.g., 0.001), and then fine-tune the model to restore the accuracy. For AlexNet and VGG-16, this method can reduce the number of weights by 9x and 13x with little loss of accuracy on ImageNet. The follow-up work "Deep Compression" [Han et al., 2015a], which blends the advantages of pruning, weight sharing and Huffman coding to compress DNNs, further pushes the compression ratio to 35–49x.

However, for energy-constrained end devices, the above magnitude-based weight pruning method may not be directly applicable, since empirical measurements show that the reduction of the number of weights does not necessarily translate into significant energy saving [Chen et al., 2016]. This is because for DNNs as exemplified by AlexNet, the energy of the convolutional layers dominates the total energy cost, while the number in the fully-connected layers contributes most of the total number of weights in the DNN. This suggests that the number of weights may not be a good indicator for energy, and the weight pruning should be directly energy-aware for end devices. As the first step toward this end, an online DNN energy estimation tool (https://energyestimation.mit.edu/) has been developed by MIT to enable fast and easy DNN energy estimation. This fine-grained tool profiles the energy for the data movement from different levels of the memory hierarchy, the number of MACs, and the data sparsity at the granularity of DNN layer. Based on this energy estimation tool, an energy-aware pruning method called EAP [Yang et al., 2017] is proposed.

Another mainstream technique for model compression is quantization. Instead of adopting the 32-bit floating point format, this technique uses a more compact format to represent layer inputs, weights, or both. Since representing a number with fewer bits reduces memory footprint and accelerates computation, quantization improves overall computation and energy efficiency. Table 6.2 surveys the most commonly adopted number representations used by DNNs, which can be categorized into 4 types: floating-point, fixed-point, exponent and binary. The above four types can be represented by the canonical number format based on IEEE 754 Standard, as illustrated in Figure 6.2. The canonical format consists of four fields: sign (S), exponent (E), fraction (F), and bias (B), as shown in Figure 6.2. Most prior proposals for quantization tune the bit-width only for a fixed number type in an ad hoc manner, which may lead to a sub-optimal result. To address this issue, the recent work of Oh et al. [2018] investigated the problem of optimal number representations at the layer granularity, in terms of finding the optimal bit-width for the canonical format based on IEEE 754 Standard. This problem is challenging due to the combinatorial explosion of feasible number formats. In response, the authors developed a portable API called number abstract data type (ADT). It enables users to declare the data to be quantized in a layer (e.g., inputs, weights, or both) as Number type. By doing so, ADT

Table 6.2: Survey of popular number representations for DNNs (— means not disclosed)

Format	Number Type	Bit Width	Network	Canonical Form $< n_s, n_e, n_f, B >$
FLOAT32		32	Default for all DNNs	$< 1, 8, 23, 127 >$
Half-precision	Floating Point	16	CNN, LSTM, DCGAN, Deep-Speech2 (Micikevicius et al., 2018)	$< 1, 5, 10, 15 >$
MS-fp9,BFP		9	DNNs on Brainwave (Chung et al., 2018; Fowers et al., 2018)	$<1, 5, 3, — >$
FIXED 16		16	CaffeNet, VGG16, VGG16-SVD (Qiu et al., 2016)	$< 1, 0, 15, 0 >$
FIXED 11	Fixed Point	10	CNN (Courbariaux et al., 2014)	$< 1, 0, 10, — >$
FIXED 8		8	MLP, CNN, LSTM (Jouppi et al., 2017)	$< 1, 0, 7, — >$
LogQuant4	Exponent	4	AlexNet, VGG16 (Miyashita et al., 2016)	$< 1, 3, 0, 3 >$
BIN	Binary	1	BinarizedNN (Courbariaux et al., 2016)	$< 1, 0, 0, 0 >$

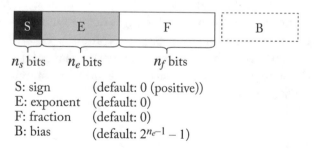

S: sign   (default: 0 (positive))
E: exponent   (default: 0)
F: fraction   (default: 0)
B: bias   (default: $2^{n_e-1} - 1$)

Figure 6.2: Canonical number format based on IEEE 754 standard.

encapsulates the internal representation of a number, thus separating the concern for developing an effective DNN from the concern of optimizing the number representation at a bit level.

Compact network architectures can be also applied to compress a DNN model. The basic idea of this method is to reduce the number of weights and operations by improving the network architecture itself, i.e., using a series of smaller filters that have fewer weights in total to replace a large filter. This approach can be applied during the network architecture design before training or after training by decomposing filters of the trained network. Specifically, when applied during the network architecture design, filters with a smaller width and height can be concatenated to emulate a larger filter. For example, MobileNets [Howard et al., 2017] and Xception use a set of 2D convolutions followed by 1x1 3D convolutions to replace a 3D convolution. More aggressively, by using many 1x1 filters to reduce the number of weights, SqueezeNet achieves an overall 50x reduction in number of weights compared to AlexNet, while maintaining the same accuracy. On the other hand, tensor decomposition can be used to decompose filters in a trained network to reduce the weights and operations. This approach first treats weights in a layer as a 4D tensor and breaks it into a combination of smaller tensors (i.e., several layers). Then, it applies low-rank approximation to further compress model. While accuracy degradation is incurred, it can be restored by fine-tuning the weights. MobileNets, Xception and SqueezeNet all take this approach to further compress the DNN models after training.

While most existing efforts use a single compression technique, they may not suffice to meet the diverse requirements and constraints on accuracy, latency, storage, and energy imposed by some IoT devices. Emerging studies have shown how different compression techniques can be coordinated to maximally compress DNN models. For example, both Deep Compression [Han et al., 2015a] and Minerva [Reagen et al., 2016] combine weight pruning and data quantization to enable fast, low-power and highly-accurate DNN inference. More recently, researchers argue that for a given DNN, the combination of compression techniques should be selected on demand, i.e., adapting to the application driven system performance (e.g., accuracy, latency and energy) and the varying resource availability across platforms (e.g., storage and processing capability). To this end, the proposed automatic optimization framework AdaDeep [Liu et al., 2018] systematically formulates the goals and constraints on accuracy, latency, storage and energy into a unified optimization problem, and leverages deep reinforcement learning (DRL) to effectively find a good combination of compression techniques. Extensive evaluations on five public datasets and across twelve mobile devices demonstrate that AdaDeep achieves up to 9.8x latency reduction, 4.3x energy efficiency improvement, and 38x storage reduction in DNNs while incurring negligible accuracy loss.

6.3.2   MODEL PARTITION

To alleviate the pressure of the edge intelligence application execution on end devices, as shown in Figure 6.3, one intuitive idea is the model partition, i.e., offloading the computational-intensive part to the edge server or the nearby mobile devices so as to obtain a better model inference performance. Model partition mainly cares about the issues of latency, energy and privacy.

The model partition can be divided into two types, namely partition between server and device and partition between devices. For the model partition between server and device, Neurosurgeon [Kang et al., 2017b] represents an iconic effort. In Neurosurgeon, the DNN model is

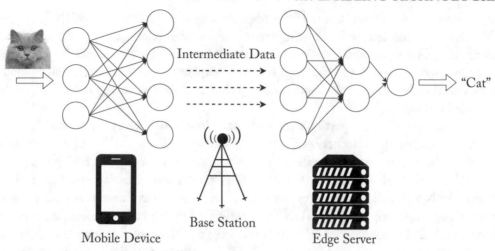

Figure 6.3: An illustration for model partition between devices and edge server.

partitioned between the device and the server. A key challenge herein is to figure out one suitable partitioning to get the optimal model inference performance. Considering from latency aspect and energy efficiency aspect, respectively, the authors propose a regression-based method to estimate the latency of each layer in the DNN model and obtain an optimal partitioning which makes the model inference meet latency requirement or energy requirement. When optimizing for best latency performance, Neurosurgeon achieves a latency speedup of 3.1x on average and up to 40.7x over the cloud-based approach. When optimizing for best energy consumption, Neurosurgeon achieves on average a 59.5% reduction in mobile energy and up to 94.7% reduction over the cloud-based approach.

Hereafter, Ko et al. [2018] propose an edge-host partitioning method, which combines model partitioning with lossy feature encoding. That is, the intermediate data after model partitioning will be compressed using lossy feature encoding before transmission. The experiment shows that the proposed approach improves the system energy efficiency and throughput by 15.3x and 16.5x compared to the edge-based method, and 2.3x and 2.5x compared to the device-based method. Also by jointly leveraging model partition and lossy feature encoding, the JALAD [Li et al., 2018c] framework formulates the model partition as an integer linear programming (ILP) problem to minimize the model inference latency under a guaranteed accuracy constraint. The simulation demonstrates that the method can speed up model inference while guaranteeing the accuracy loss within a user-defined requirement. For DNNs those are characterized by a directed acyclic graph (DAG) rather than a chain, optimizing the model partition to minimize the latency is proven to be NP-hard in general. In response, Huang et al. [2019] propose an approximation algorithm that provides worst-case performance guarantee, based on the graph min-cut method. The above frameworks are all based on an assumption that the server has the DNN model of the edge intelligence application. IONN [Jeong et al., 2018b] propose an incremental offloading technique for edge intelligence applications. IONN partitions the DNN layers and incrementally uploads them to allow collaborative DNN model inference by mobile devices and the edge server. Compared to the approach that uploads the entire model, IONN significantly improves query performance and energy consumption during DNN model uploading.

Another type of model partitioning is the partitioning between devices. As the pioneering effort of model partitioning between devices, MoDNN [Mao et al., 2017a] introduces WiFi Direct technique to build a micro-scale computing cluster in WLAN with multiple authorized WiFi-enabled mobile devices for partitioned DNN model inference. The mobile device that carries the DNN task will be the group owner and the others act as the worker nodes. Two partitioning schemes are proposed in MoDNN to accelerate DNN layer execution. The experiment shows that with 2–4 worker nodes, MoDNN accelerates DNN model inference by 2.17–4.28x. In the follow-up work MeDNN [Mao et al., 2017b], a greedy two-dimensional partitioning is proposed to adaptively partition DNN model onto multiple mobile devices and utilize a structured sparsity pruning technique to compress the DNN model. MeDNN improves DNN model inference by 1.86–2.44x with 2–4 worker nodes and saves 26.5% of additional computing time and 14.2% of extra communication time. Note that DNN layers are partitioned horizontally in MoDNN and MeDNN. In contrast, DeepThings [Zhao et al., 2018b] employs a fused tile partitioning method that partitions the DNN layers vertically to reduce the memory footprint. The Fused Tile partitioning method can reduce memory usage by more than 68% without accuracy loss and DeepThings speedups DNN model inference of 1.7x–3.5x on 2–6 edge devices with no more than 23 MB memory each.

DeepX [Lane et al., 2016] also aims to partition DNN models but it only partitions the DNN model into several sub-models and distributes them on local processors. DeepX proposes two schemes: runtime Layer Compression (RLC) and Deep Architecture Decomposition (DAD). The layer after compression will be executed by specific local processors (CPU, GPU, and DSP). Note that the scheduler needs to be optimized when we have multiple tasks of model partition. LEO [Georgiev et al., 2016] is a novel sensing algorithm scheduler that maximizes the performance for multiple continuous mobile sensor applications by partitioning the sensing algorithm execution and distributing tasks on CPU, co-processor, GPU and the cloud. And OpenVDAP [Zhang et al., 2018b] proposes a heterogeneous vehicle computing platform, including a tasks scheduling framework. The framework schedules the tasks to specific acceleration hardware based on task computing characteristics and hardware utilization.

Model Early-Exit

A DNN model with a high accuracy usually has a deep structure, and it consumes a large amount of resources to execute such a DNN model on the end device. To accelerate model inference, model early-exit method leverages output data of early layer to get the classification result, which

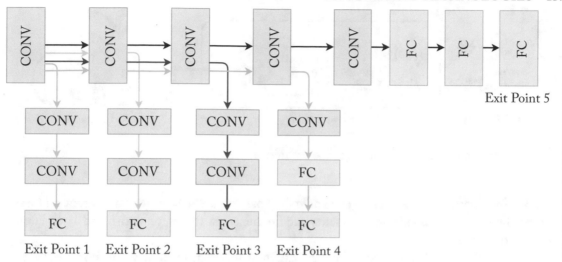

Exit Point 5

Exit Point 1   Exit Point 2   Exit Point 3   Exit Point 4

Figure 6.4: A CNN model with five exit points.

means that the inference process is finished by using a partial DNN model. Latency is the optimization target of model early-exit.

BranchyNet [Teerapittayanon et al., 2016] is a programming framework that implements the model early-exit mechanism. With BranchyNet, the standard DNN model structure is modified by adding exit branches at certain layer locations. Each exit branch is an exit point and shares part of DNN layers with the standard DNN model. Figure 6.4 shows a CNN model with three five points. The input data can be classified at these diverse early exit points. Since the BranchyNet model can be viewed as a DNN model with multi-tasks, the training of BranchyNet is a joint optimization problem. For the trained BranchyNet model, the mobile devices can execute the branches at early exit points, if the accuracy cannot meet a requirement, and the intermediate data of the early exit point will be sent to the edge server and executed by the later exit point. The experiments show that BranchyNet speedups 2x–6x on both CPU and GPU.

Based on BranchyNet, a framework named DDNNs [Teerapittayanon et al., 2017] for distributed deep neural networks across the cloud, edge and devices is proposed. DDNNs has a three-layer structure framework, including device layer, edge server layer and cloud layer. Each layer represents an exit point of BranchyNet. Three aggregation methods including max pooling (MP), average pooling (AP) and concatenation (CC) are proposed. The aggregation methods work when multiple mobile devices send intermediate to an edge server or when multiple edge servers send intermediate data to the cloud data center. MP aggregates the data vectors by taking the max of each component. AP aggregates the data vectors by taking the average of each component. CC just simply concatenates the data vectors as one vector. Also built on top of BranchyNet, Edgent [Li et al., 2018b] is proposed to navigate the accuracy-latency tradeoff

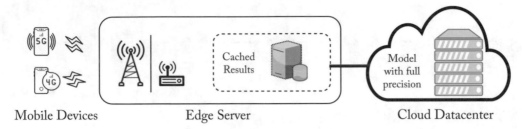

Figure 6.5: The process of semantic cache technique.

when jointly applying model early-exit and model partition. The basic idea of Edgent is to maximize the accuracy under a given latency requirement, via the regression-based layer latency prediction model.

In addition to BranchyNet, there are different methods to implement model early-exit. For example, Cascading network [Leroux et al., 2017] simply adds max pooling layer and fully-connected layer to the standard DNN model and achieves a speedup of 20%. DeepIns [Li et al., 2018d] proposes a manufacture inspection system for the smart industry using DNN model early-exit. In DeepIns, edge devices are responsible for data collection, while the edge server acts as the first exit point and the cloud data center acts as the second exit point. Then Lo et al. [2017] propose to add an authentic operation (AO) unit to the basic BranchyNet model. The AO unit determines whether an input has to be transferred to the edge server or cloud data center for further execution by setting different threshold criteria of confidence level for different DNN model output classes, and Bolukbasi et al. [2017] train a policy that determines whether the current samples should proceed to the next layer by adding regularization to the evaluation latency of the DNN model.

6.3.3   EDGE CACHING

Edge caching is a new kind of methods used to accelerate DNN model inference, i.e., optimizing the latency issue, by caching the DNN inference results. The core idea of edge caching is to cache and reuse the task results as the prediction of image classification at the network edge, hence reducing the querying latency of edge intelligence application. Figure 6.5 shows the basic process of the semantic cache technique: if the request from mobile devices hits the cached results stored in the edge server, the edge server will return the result; otherwise, the request will be transferred to the cloud data center for inference with the model of full precision.

Glimpse [Chen et al., 2015b] is a pioneering effort to introduce a cache technique to DNN inference task which achieves an acceleration of 1.6–5.5x. For an object detection application, Glimpse proposes to reuse the stale detection result to detect the object on current frames. The results of the detected object of the stale frames are cached on the mobile devices, and then Glimpse extracts a subset of these cached results and computes the optical flow of features between the processed frames and the current frame. The computing results of optical flow will guide us to move the bounding box to the right location in the current frame.

But caching results locally does not scale beyond tens of images. Cachier [Drolia et al., 2017b] is proposed to achieve recognition of thousands of objects. In Cachier, results of edge intelligence application are cached in the edge server, storing the features of input (e.g., image) and the corresponding task results. Then Cachier uses the least frequently used (LFS) as the cache replacement strategy. If the input cannot hit the cache, the edge server will transfer the input to the cloud data center. Cachier can increase responsiveness by 3x or more. Precog [Drolia et al., 2017a] is the extension of Cachier. In Precog, the cached data is not only stored in the edge server but also in the mobile device. Precog uses predictions of Markov chains to prefetch data onto the mobile device and reaches a speedup of 5x. In addition, Precog also proposes to dynamically adjust the cached feature extraction model on the mobile device according to the environment information. Shadow Puppets is another improved version of Cachier. Cachier extracts features from input using the standard feature extraction like locality sensitive hashing (LSH), but these features may not reflect the similarity as precise as the human dose. Then in Shadow Puppets [Venugopal et al., 2018], it uses a small-footprint DNN to generate hash codes to represent the input data and get a remarkable latency improvement of 5–10x.

Considering the application scenario that the same application runs on multiple devices in close proximity and the DNN model often processes similar input data, FoggyCache [Guo et al., 2018] is proposed to minimize the redundant computations. To address two challenges, i.e., (1) how to index the input data with a constant lookup quality since the input data distribution is unknown, (2) how to represent the similarity of the input data, FoggyCache proposes adaptive locality sensitive hashing (A-LSH) and homogenized kNN (H-kNN) schemes, respectively. FoggyCache reduces computation latency and energy consumption by a factor of 3–10x.

6.3.4   INPUT FILTERING

Input filtering is an efficient method to accelerate DNN model inference, especially for the video analytics. s shown in Figure 6.6, the key idea of input filtering is to remove the non-target-object frames of input data to avoid redundant computation of DNN model inference, hence improving inference accuracy, shortening inference latency and reducing energy cost.

NoScope [Kang et al., 2017a] is proposed to accelerate video analysis by skipping the frames that have little change. To this end, NoScope implements a difference detector that highlights temporal differences across frames. For example, the detector monitors the frames to check whether cars appear in the frames and the frame with cars will be processed in DNN model inference. The difference is detected by using lightweight binary classifiers. Under a scenario of continuous video transmission from a swarm of drones, Wang et al. [2018a] optimize for the first hop wireless bandwidth of DNN inference. In particular, four strategies are proposed to reduce total transmission: EarlyDiscard, Just-in-Time-Learning (JITL), Reachback and Context-Aware.

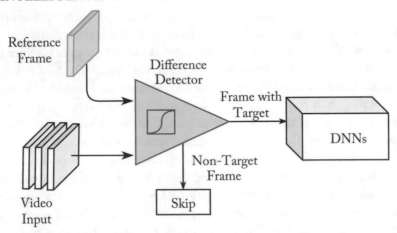

Figure 6.6: The workflow of input filtering.

Wang et al. [2018a] optimize for the first hop wireless bandwidth of DNN inference. Under a scenario of continuous video transmission from a swarm of drones, Wang at el. propose four different strategies to reduce total transmission: EarlyDiscard, Just-in-Time-Learning (JITL), Reachback and Context-Aware. EarlyDiscard is a content-based filtering method. It uses a trained lightweight DNN like MobileNet (classification task) to filter-out uninteresting frames. To address the problem that the EarlyDiscard has a low accuracy due to the lightweight classifier, JITL adds a cascade filter to reduce false-positive frames transmitted by EarlyDiscard. The cascade filter is trained periodically and pushed to the drone. The Reachback is that the edge server can fetch any filtered frames by previous two strategies from the drone on-demand and the Context-Aware performs task-specific optimizations such as detecting pixels with certain colors.

FFS-VA [Zhang et al., 2018a] is a pipelined system for multi-stage video analytic. There are three stages to build the filtering system of FFS-VA. The first is a stream-specialized different detector (SDD) which is used to remove the frames only containing a background. The second is a stream-specialized network model (SNM) to identify target-object frames. And the third is a Tiny-YOLO-Voc (T-YOLO) model to remove the frames whose target objects are fewer than a threshold. Canel et al. [2018] propose a two-stage filtering system for video analytics. It first extracts the semantic content of the frames by outputting the intermediate data of DNN, and then these output features are accumulated in a frame buffer. The buffer is viewed as a directed acyclic graph and the filtering system uses Euclidean distance as the similarity metric to figure out top-k interesting frames.

The above frameworks focus on filtering uninteresting frames of a video stream for a single camera. ReXCam [Jain et al., 2018] is further proposed to accelerate DNN model inference on cross-camera analytics by leveraging a learned spatiotemporal model to filter video frames.

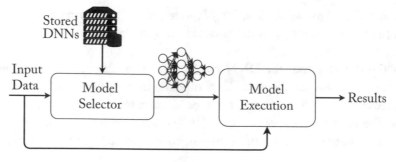

Figure 6.7: The process of DNN model selection.

Moreover, ReXCam reduces computation workload by 4.6x and improves DNN model inference accuracy by 27%.

6.3.5   MODEL SELECTION

The model selection method is proposed to optimize the DNN inference issue of latency, accuracy and energy. The basic idea of model selection is to first train a set of DNN models for the same task with various model size offline, and then adaptively select the model for inference online. Model selection is similar to the model early-exit, in the sense that the exit point of model early-exit mechanism can be viewed as a DNN model, with the key difference being that the exit point shares part of DNN layers with the main branch model but the models in the model selection mechanism are independent.

Park et al. [2015] propose a big/little DNN model selection framework, where a little and fast model is used to classify the input data while a big model is only used when the confidence of the little model is less than a predefined threshold. Taylor et al. [2018] point out that different DNN models (e.g., MobileNet, ResNet, Inception) reach lowest inference latency or highest accuracy on different evaluation metrics (top-1 or top-5) for different images. Thus, inspired, they propose a framework for selecting the best DNN in terms of latency and accuracy, where a model selector is trained to select the best DNN for different input images. Similarly, IF-CNN [Shu et al., 2018] also trains a model selector called recognition predictor (RP) to change the model used in the task. RP is a DNN model of multi-task, meaning that RP has multiple outputs. The output of RP represents the probability of top-1 label of each candidate DNN model. The input of RP is the image, and the corresponding DNN model will be selected if the output of RP is beyond the predefined threshold.

Considering the DNN models selection according to the input data, Chameleon [Jiang et al., 2018b] propose a periodic update method to change the DNN model used for video analytics. In Chameleon, by taking the cross-video correlation into consideration, the related video sources (e.g., cameras) are grouped, and then the leader of the group will search for top-k best configuration and share the top-k configuration to its followers in the same group. A configuration includes frame resolution, DNN model, etc. Chameleon achieves 20–50% higher accuracy with the same amount of resources or achieves the same accuracy with only 30–50% of the resources.

Besides the optimization for DNN model inference latency, aiming at energy saving, Stamoulis et al. [2018] cast the adaptive DNN model selection issue as a hyper-parameter optimization problem by taking into account the accuracy and communication constraints imposed by the devices. Then a Bayesian Optimization (BO) is adopted to solve this problem, achieving an improvement by up to 6x in terms of minimum energy per image under accuracy constraints.

6.3.6   SUPPORT FOR MULTI-TENANCY

In practice, an end or edge device typically runs more than one DNN applications concurrently. For example, the advanced driver assistance system (ADAS) for internet vehicles simultaneously runs DNN programs for vehicle detection, pedestrian detection, traffic sign recognition, and lane line detection. In this case, multiple DNN applications would compete for the limited resource. Without careful support for multi-tenancy, i.e., resource allocation and task scheduling for those concurrent applications, the global efficiency would be greatly deteriorated. The support for multi-tenancy focuses on the optimization of energy and memory footprint.

Taking the dynamics of runtime resources into consideration, NestDNN [Fang et al., 2018a] is proposed to achieve flexible resource-accuracy trade-offs for each DNN model. NestDNN addresses two challenges: (1) how to enable a DNN model to provide flexible resources-accuracy trade-offs; (2) how to balance a resource-accuracy trade-off for each concurrent DNN model. NestDNN implements a new model pruning and recovery scheme, by transforming the DNN model into a single compact multi-capacity model which consists of a set of descendent models. Each descendent model offers a unique resource-accuracy trade-off. For each concurrent descendent model, NestDNN encodes its accuracy and latency into a cost function, based on which a resource-accuracy runtime scheduler is built to achieve the optimal trade-off for each concurrent descendent model. NestDNN obtains an improvement of 4.2% in inference accuracy, 2x in video frame processing rate and a reduction of 1.7x in energy consumption.

To address the challenge in enabling flexible trade-offs, Mainstream [Jiang et al., 2018a] uses the popular transfer-learning DNN training method to train multiple DNN models with different degrees of accuracy and implements a greedy approach to find the optimal scheduler that fits the cost budget. For multiple DNN model executions on one single device, Hive-Mind [Narayanan et al., 2018] is proposed to improve the GPU utilization for these concurrent workloads. HiveMind consists of two key components: a compiler and a runtime module. The compiler optimizes the data transmission, data preprocessing and computation across the workloads, and then the runtime module transforms the optimized models into an execution DAG, which will be executed on the GPU while trying to extract as much concurrency as possible.

At a finer granularity, DeepEye [Mathur et al., 2017] is proposed to optimizing the inference of multi-task on the mobile device by scheduling the executions of heterogeneous DNN layers. DeepEye first segregates DNN layers of all task into two pools: convolution layers and fully-connected layers. For the convolution layers, a FIFO queue based execution strategy is employed. For the fully-connected layers, DeepEye adopts a greedy approach for caching the parameters of the fully connected layers to maximize memory utilization.

Different from the above framework optimizing for multi-task on the local device, Chameleon [Jiang et al., 2018b] optimizes the profiling cost for multiple video processing. In Chameleon, the configuration (e.g., frame resolution, the selected DNN model) of video analytic tasks needs to be profiled periodically. Considering the spatial correlation between different video sources, cameras with similar characteristics are grouped, among which the leader will search for the best top-k configuration and share them to the followers. Consequently, the cost of profiling is amortized across similar cameras with minimal reduction in accuracy.

Besides, the optimization for the edge-device scenario is also considered. Note that mobile clients connect to the edge server and the edge server connects to the backend cluster. VideoEdge [Hung et al., 2018] is one framework to optimize the video query to maximize the average query accuracy within the available resources. In VideoEdge, hierarchical clusters are formed with multiple computing capabilities and network bandwidth. Then, VideoEdge dynamically places processing modules of a query across these clusters and selects one configuration for the query to maximize the average accuracy, which can be formulated as a multiple-choice multi-dimensional knapsack problem and solved using a greedy heuristic method. VideoEdge achieves an improvement of 25.4x in accuracy compared to a fair allocation and 5.4x in accuracy compared to another video query system (VideoStorm). ATOMS is another framework to optimize the offloading for the edge-device scenario. In ATOMS [Fang et al., 2018b], a Plan-Schedule strategy is proposed, which consists of a planning phase and a scheduling phase. In the planning phase, ATOMS predicts time slots of future tasks from all mobile clients connected to the edge server, detects contention, coordinates tasks, and informs clients about new offloading times. In the scheduling phase, for each arriving task, ATOMS selects the machine that has minimal estimated processor contention to execute it.

6.3.7   APPLICATION-SPECIFIC OPTIMIZATION

While the above optimizing techniques are generally applicable to edge intelligence applications, application-specific optimization can be exploited to further optimize the performance of EI applications, i.e., accuracy, latency, energy, and memory footprint. For example, for video-based applications, two knobs, i.e., frame rate and resolution can be flexibly adjusted to reduce resource demand. However, since such resource-sensitive knobs also deteriorate the inference accuracy, they naturally incur a cost-accuracy tradeoff. This requires us to strike a nice balance between the resource cost and inference accuracy when tuning the video frame rate and resolution.

Toward the above goal, Chameleon [Jiang et al., 2018b] adjusts the knobs for each video analytic task by sharing the best top-k configuration between each task. In Chameleon, the video tasks are grouped according to the spatial correlation, and then the leader of each group will search for the best top-k configurations and share them with the followers. DeepDecision [Ran et al., 2018] formulates the knob-tuning problem as a multiple-choice multiple-constraint knapsack program and solves it with an improved brute-force search method.

Application-level optimization improves the performance of edge intelligence applications using application-specific optimizing scheme. For example, for the video-based applications, several efficient pipelines have been proposed to improve the performance of DNN model inference in terms of accuracy and latency. VideoStorm [Zhang et al., 2017a] is a framework to maximize the accuracy and minimize the latency of edge intelligence application within the resource capacity by profiling the task configuration (e.g., frame rate, resolution) and resource allocation (e.g., CPU). Different from VideoStorm profiling the configuration for each running query, Chameleon [Jiang et al., 2018b] adjusts the configuration for each video analytic task by sharing the best top-k configuration between each task. In Chameleon, the video tasks are grouped according to the spatial correlation, and then the leader of the group will search for the best top-k configurations and share them with the followers.

Also considering adjusting the configuration of DNN model inference for video analytic (e.g., AR application), DeepDecision [Ran et al., 2018] addresses the challenge in another scenario that a tiny DNN model and a big DNN model are stored in the mobile device and a big DNN model is in the edge server. In addition to profiling the configuration, DeepDecision needs to consider when to offload data to the server. The goal of DeepDecision is to maximize two key metrics: frame rate, the number of processed frame per second, and accuracy. To this end, different from the above methods used to profile the configuration of DNN inference, Grulich and Nawab [2018] propose to combine different accelerating methods of DNN model inference to achieve inference performance.

It is also worth noting that, in the computer architecture community, hardware acceleration for efficient DNN inference has been a very hot topic and gathered extensive research efforts. Interested readers are encouraged to refer to the recent monograph [Sze et al., 2017] for more comprehensive discussions about recent advancements on hardware acceleration for DNN processing.

6.4   SUMMARY

To showcase the applications of the above enabling techniques for edge intelligence model inference, the relevant systems and frameworks are summarized in Table 6.3, including the perspectives of target applications, architecture and EI level, optimization objectives, and adopted techniques, as well as effectiveness.

Clearly, existing systems and frameworks have adopted different subsets of enabling techniques tailored to specific edge intelligence applications and requirements. To maximize the

Table 6.3: An overview of systems and frameworks on EI model inference (*Continues*).

System or Framework	Application	Architecture	EI Level	Objectives	Optimization Technology	Online/ Offline	Effectiveness
VideoEdge (Hung et al., 2018)	Video Analytics	Cloud-Edge-Device	Level-1	• Accuracy • Resource Cost	• Frame rate adaptation • Resolution adaptation • Multi-tenancy • Service placement	Online	• Accuracy improvement: 5.4–25.4×
Chameleon (Jiang et al., 2018b)	Video Analytics	Device-Cloud	Level-1	• Accuracy • Resource Cost	• Frame rate adaptation • Resolution adaptation • Model selection	Online	• Resource reduction: 2–3×
DeepX (Lane et al., 2016)	Mobile Sensing App	On Device	Level-2	• Accuracy • Energy	• Model compression • Model partition	Online	• Energy reduction: 7.12–26.7×
Edgent (Li et al., 2018b)	N/A	Device-Edge	Level-2	• Accuracy • Latency	• Early-exit • Model partition	Offline	• Accuracy improvement
AdaDeep (Liu et al., 2018)	N/A	On Device	Level-3	• Accuracy • Energy • Storage	• Model compression	Online	• Latency reduction: 9.8× • Energy reduction: 4.3× • Storage reduction: 38×
DeepIns (Li et al., 2018d)	IIoT	Edge-Cloud	Level-1	• Accuracy • Latency	• Early-exit	Offline	• Latency reduction: 0.98–1.21×
Neurosurgeon (Kang et al., 2017b)	N/A	Device-Cloud	Level-1	• Latency • Energy	• Model partition	N/A	• Latency reduction: 3.1–40.7× • Energy reduction: 59.5%–94.7%
Minerva (Reagen et al., 2016)	N/A	On-Device	Level-3	• Energy	• Hardware Acceleration • Model Compression	Offline	• Energy saving: 8×

Table 6.3: (*Continued*). An overview of systems and frameworks on EI Model Inference

						Online	
FoggyCache (Guo et al., 2018)	IIoT	Device-Edge	Level-2	• Accuracy • Latency	• Edge Caching	Online	• Latency reduction: 3–10× • Energy reduction: 3–10×
NoScope (Kang et al., 2017a)	Video Analytics	Cloud	Cloud Intelligence	• Latency	• Input filtering	N/A	• Latency reduction: 265–15,500×
JALAD (Li et al., 2018c)	N/A	Device-Cloud	Level-1	• Latency	• Model compression • Model partition	Offline	• Latency reduction: 1–25.1×
DDNNs (Teerapittayanon et al., 2017)	N/A	Cloud-Edge-Device	Level-1	• Latency • Accuracy	• Model selection	N/A	• Latency reduction: over 20×
FFS-VA (Zhang et al., 2018a)	Video Analytics	On-Device	Level-3	• Latency	• Input filtering • Multi-tenancy	N/A	• Latency reduction: 3× • Throughput improvement: more than 7×
Cachier (Drolia et al., 2017b)	N/A	Cloud-Edge	Level-1	• Through-put	• Edge Caching	N/A	• Throughput improvement: more than 3×
Premodel (Taylor et al., 2018)	N/A	On-Device	Level-3	• Accuracy • Latency	• Input filtering • Model selection	N/A	• Accuracy improvement: 7.52% • Latency reduction: 1.8×
DeepDecision (Ran et al., 2018)	Video Analytics	Cloud-Edge	Level-1	• Accuracy • Latency • Energy	• Application-level optimization • Model selection	N/A	• Latency reduction: 2–10×

overall performance of a generic edge intelligence system, comprehensive enabling techniques and various optimization methods should work in a cooperative manner to provide rich design flexibility. Nevertheless, we would face a high dimensional configuration problem that requires determining a large number of performance-critical configuration parameters in real time. Taking video analytics for an example, the high-dimensional configuration parameters can include video frame rate, resolution, model selection and model early-exit, etc. Due to the combinatorial nature, high-dimensional configuration problem involves a huge search space of parameters and is very challenging to tackle.

CHAPTER 7

On-Demand Accelerating Deep Neural Network Inference via Edge Computing

In this chapter, we study how to accelerate DNN inference under device-edge synergy, by jointly applying the two knobs of DNN model partitioning and right-sizing.

7.1 INTRODUCTION

Recall that in Chapter 6 we reviewed the architectures and enabling techniques for effective model inference at the edge side. In this chapter, we present how to coordinate various enabling techniques in a practical dynamical edge computing environment, and thus fully unleash the potential of EI model inference. Specifically, we propose Edgent, a low-latency co-inference framework via device-edge synergy. Toward low-latency edge intelligence, Edgent pursues two design knobs. The first is DNN partitioning, which adaptively partitions DNN computation between mobile devices and the edge server according to the available bandwidth so as to utilize the computation capability of the edge server. However, it is insufficient to meet the stringent responsiveness requirement of some mission-critical applications since the execution performance is still restrained by the rest of the model running on mobile devices. Therefore, Edgent further integrates the second knob, DNN right-sizing, which accelerates DNN inference by early exiting inference at an intermediate DNN layer. Essentially, the early-exit mechanism involves a latency-accuracy tradeoff. To strike a balance on the tradeoff with the existing resources, Edgent makes a joint optimization on both DNN partitioning and DNN right-sizing in an on-demand manner. Specifically, for mission-critical applications that are typically with a pre-defined latency requirement, Edgent maximizes the accuracy while promising the latency requirement.

We consider the design of Edgent under the versatile network conditions in practical deployment in both static network environment and dynamic network environment. Specifically, in a static network environment (e.g., local area network with fiber connection), we regard the bandwidth as stable and figure out a collaboration strategy through execution latency estimation based on the current bandwidth. In this case, Edgent trains regression models to predict the layer-wise inference latency and accordingly derives the optimal configurations for DNN partitioning and DNN right-sizing. In a dynamic network environment (e.g., 5G cellular network, vehicular network), to alleviate the impact of network fluctuation, we build a look-up table by profiling and recording the optimal strategy selection for each bandwidth state and design a runtime optimizer to detect the bandwidth state transition and generate the optimal strategy accordingly. Extensive evaluations based on prototype implementation demonstrate Edgent's effectiveness in enabling on-demand low-latency edge intelligence. To summarize, the contributions of this chapter are as follows.

- We propose Edgent, a framework for on-demand DNN collaborative inference through device-edge synergy, in which we jointly optimize DNN partitioning and DNN right-sizing to maximize the inference accuracy while promising application latency requirement.

- We consider the design of Edgent in both static network environment and dynamic network environment, and accordingly put forward regression model based and bandwidth state detection based optimal strategy selection schemes for both cases.

- We implement and experiment Edgent prototype with a Raspberry Pi as the end device. The evaluation results based on the real-world network trace dataset demonstrate the effectiveness of the proposed Edgent framework.

The rest of this chapter is organized as follows. First, we present the background and motivation in Section 7.2. Then we propose the design of Edgent in Section 7.3. The results of the performance evaluation are shown in Section 7.4 to demonstrate the effectiveness of Edgent. Finally, we conclude in Section 7.5.

7.2   BACKGROUND AND MOTIVATION

In this section, we first analyze the limitation of edge- and device-only methods, motivated by which we explore the way to utilize DNN partitioning and right-sizing to accelerate DNN inference with device-edge synergy.

7.2.1   INSUFFICIENCY OF DEVICE-ONLY OR EDGE-ONLY DNN INFERENCE

Traditional mobile DNN computation is either solely performed on mobile devices or wholly offloaded to cloud/edge servers. Unfortunately, both approaches may lead to poor performance (i.e., high end-to-end latency), making it difficult to meet real-time applications latency requirements [Chen et al., 2018b]. For illustration, we employ a Raspberry Pi and a desktop PC to emulate the mobile device and edge server, respectively, and perform image recognition task over cifar-10 dataset with the classical AlexNet model [Krizhevsky et al., 2012]. Figure 7.1 depicts the end-to-end latency subdivision of different methods under different bandwidths on the edge server and the mobile device (simplified as Edge and Device in Figure 7.1). As shown in Figure 7.1, it takes more than 2 s to finish the inference task on a resource-constrained mobile

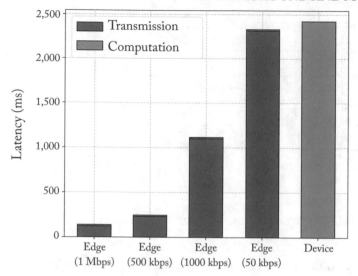

Figure 7.1: Execution runtime of edge-only (Edge) and device-only (Device) approaches of AlexNet under different bandwidths.

device. As a contrast, the edge server only spends 0.123 s for inference under a 1 Mbps bandwidth. However, as the bandwidth drops, the execution latency of the edge-only method climbs rapidly (the latency climbs to 2.317 s when the bandwidth drops to 50 kbps). This indicates that the performance of the edge-only method is dominated by the data transmission latency (the computation time at server side remains at $\sim$ 10 ms) and is therefore highly sensitive to the available bandwidth. Considering the scarcity of network bandwidth resources in practice (e.g., due to network resource contention between users and applications) and the limitation of computation resources on mobile devices, device- and edge-only methods are insufficient to support emerging mobile applications in stringent real-time requirements.

7.2.2   DNN PARTITIONING AND RIGHT-SIZING TOWARD EDGE INTELLIGENCE

DNN Partitioning: To better understand the performance bottleneck of DNN inference, in Figure 7.2 we refine the layer-wise execution latency (on Raspberry Pi) and the intermediate output data size per layer. Seen from Figure 7.2, the latency and output data size of each layer show great heterogeneity, implying that a layer with a higher latency may not output larger data size. Based on this observation, an intuitive idea is DNN partitioning, i.e., dividing DNN into two parts and offloading the computation-intensive part to the server at a low transmission cost and thus reducing total end-to-end execution latency. As an illustration, we select the second local response normalization layer (i.e., lrn_2) in Figure 7.2 as the partition point and the layers

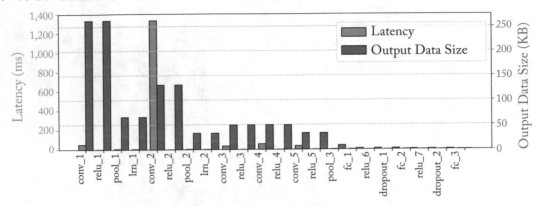

Figure 7.2: AlexNet layer-wise runtime and output data size on Raspberry Pi.

before the point are offloaded to the server side while the rest remains on the device. Through model partitioning between the device and edge, hybrid computation resources in proximity can be in comprehensive utilization toward low-latency DNN inference.

DNN Right-Sizing: Although DNN partitioning can significantly reduce the latency, it should be noted that with the optimal DNN partitioning, the inference latency is still constrained by the remaining computation on the mobile device. To further reduce the execution latency, the DNN right-sizing method is employed in conjunction with DNN partitioning. DNN right-sizing accelerates DNN inference through the early-exit mechanism. For example, by training DNN models with multiple exit points, a standard AlexNet model can be derived as a branchy AlexNet as Figure 7.3 shows, where a shorter branch (e.g., the branch ended with the exit point (1) implies a smaller model size and thus a shorter runtime. Note that in Figure 7.3 only the convolutional layers (CONV) and the fully connected layers (FC) are drawn for the ease of illustration. This novel branchy structure demands novel training methods. In this chapter, we implement the branchy model training with the assist of the open-source BranchyNet [Teerapittayanon et al., 2016] framework.

Problem Definition: Clearly, DNN right-sizing leads to a latency-accuracy tradeoff, i.e., while the early-exit mechanism reduces the total inference latency, it hurts the inference accuracy. Considering the fact that some latency-sensitive applications have strict deadlines but can tolerate moderate precision losses, we can strike a good balance between latency and accuracy in an on-demand manner. More precisely, the problem addressed in this chapter can be summarized as how to make a joint optimization on DNN partitioning and DNN right-sizing in order to maximize the inference accuracy without violating the pre-defined latency requirement.

Formally, we define the problem as follows. For a DNN model with M exit points, we denote that the branch of ith exit point has N_i layers and D_p is the output data size of the pth layer. We also denote the execution latency ED_j of the jth layer running on the device and the

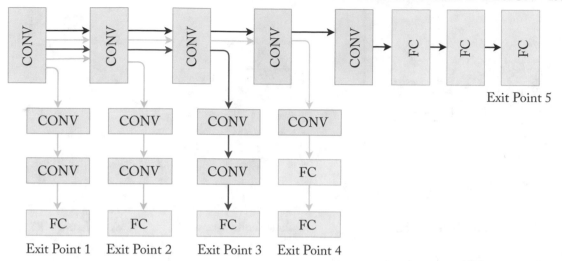

Figure 7.3: A branchy AlexNet model with five exit points.

latency ES_j of the j th layer running on the server. Given the current network bandwidth B, the objective is to find the proper pair of exit point i and partition point p to maximize the DNN model inference accuracy $ACC_{i,p}$ subject to the maximum latency constraint L, i.e.,

$$\max_{i \in \{1,...,M\}, p \in \{1,...,N_i\}} ACC_{i,p}$$

$$\text{subject to} \quad \sum_{j=1}^{p-1} ES_j + \sum_{j=p}^{N_i} ED_j + \frac{D_{p-1}}{B} \le L. \tag{7.1}$$

7.3   FRAMEWORK AND DESIGN

In this section, we present the design of Edgent, which can generate the optimal collaborative DNN inference strategy in run-time that maximizes the accuracy while meeting the latency requirement in both the static and dynamic bandwidth environments.

7.3.1   FRAMEWORK OVERVIEW

As illustrated in Figure 7.4, the workflows of Edgent mainly consist of three stages: offline configuration stage, online tuning stage, and co-inference stage.

At the **offline configuration stage**, Edgent inputs the DNN model to the *Static/Dynamic Configurator* component and obtains the corresponding configuration for online tuning. To be specific, composed of trained regression models and branchy DNN model, the static configu-

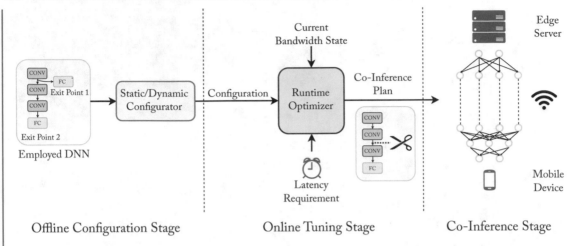

Figure 7.4: Edgent framework overview.

ration will be employed where the bandwidth keeps stable during the DNN inference (which will be detailed in Section 7.3.2), while composed of the trained branchy DNN and the optimal selection for different bandwidth states, the dynamic configuration will be used adaptive to the state dynamics (which will be detailed in Section 7.3.3).

At the **online tuning stage**, Edgent measures the current bandwidth state and makes a joint optimization on DNN partitioning and DNN right-sizing based on the given latency requirement and the configuration obtained offline, aiming at maximizing the inference accuracy under the given latency requirement.

At the **co-inference stage**, based on the co-inference plan (i.e., the selected exit point and partition point) generated at the online tuning stage, the layers before the partition point will be executed on the edge server with the rest remaining on the device (the other way around can be also supported when the data samples are input from the device).

During DNN inference, the bandwidth between the mobile device and the edge server may be relatively stable or frequently changing. Though Edgent runs on the same workflow in both static and dynamic network environments, the function of *Configurator* component and *Runtime Optimizer* component differ. Specifically, under a static bandwidth environment, the configurator trains regression models to predict inference latency and the branchy DNN to enable early-exit mechanism. The *Static Configurator* includes the regression models and the branchy DNN that are grained offline, and the *Runtime Optimizer* will compute the optimal co-inference plan in the online manner. Under a dynamic bandwidth environment, the *Dynamic Configurator* creates a configuration map that records the optimal selection for current bandwidth state via the change point detector, which will then be input to the *Runtime Optimizer* to pick up the optimal co-inference plan accordingly. In the following, we will discuss

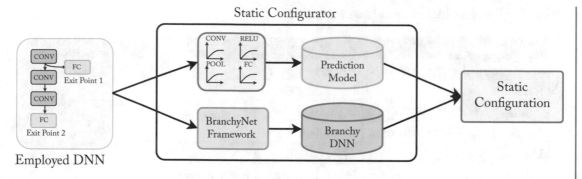

Figure 7.5: The static configurator of Edgent.

the design details of the configurator and optimizer for the static and dynamic environments, respectively.

7.3.2 EDGENT FOR STATIC ENVIRONMENTS

As a starting point, we first consider our framework design in the case of the static network environment. The key idea of the static configurator is to train regression models to predict the layer-wise inference latency and train the branchy model to enable early-exit mechanism. The configurator specialized for the static bandwidth environment is shown in Figure 7.5.

At the **offline configuration stage**, the static configurator initiates two tasks: (1) profile layer-wise inference latency on the mobile device and the edge server, respectively, and accordingly train regression models for different kind of DNN layer (e.g., Convolution, Fully-Connected, etc.); and (2) train the DNN model with multiple exit points via BranchyNet framework to obtain branchy DNN. The profiling process is to record the inference latency of each type of layers rather than that of the entire model. Based on the profiling results, we establish the prediction model for each type of layers by performing a series of regression models with independent variables shown in Table 7.1. Since there are limited types of layers in a typical DNN, the profiling cost is moderate. Since the layer-wise inference latency is infrastructure-dependent, for a DNN inference task Edgent only needs to conduct the above two tasks for once at the initialization.

At the **online tuning stage**, using the static configuration (i.e., the prediction model and the branchy DNN), the *Runtime Optimizer* component searches the optimal exit point and partition point to maximize the accuracy while ensuring the execution deadline with three inputs: (1) the static configuration, (2) the measured bandwidth between the edge server and the end device, and (3) the latency requirement. The search process of joint optimization on the selection of partition point and exit point is described in Algorithm 7.6. For a DNN model with M exit points, recall that the branch of ith exit point has N_i layers and D_p is the output of the

Table 7.1: The variables of prediction models

Type of DNN Layer	Variable(s)
Convolutional	# of input feature maps, (filter size/stride)^2×(# of filters)
Relu	Input data size
Pooling	Input data size, output data size
Local Response Normalization	Input data size
Dropout	Input data size
Fully Connected	Input data size, output data size

pth layer. We use the above-mentioned regression models to predict the latency ED_j of the jth layer running on the device and the latency ES_j of the jth layer running on the server. Under a certain bandwidth B, we can calculate the total latency $T_{i,p}$ by summing up the computation latency on each side and the communication latency for transferring input data and intermediate execution result. We denote pth layer as the partition point of the branch with the ith exit point. Therefore, $p = 1$ indicates that the total inference process will only be executed on the device side (i.e., $ES_p = 0$, $D_{p-1} = 0$) whereas $p = N_i$ means the total computation is only done on the server side (i.e., $ED_p = 0$). Through the exhaustive search on different points with a linear complexity, we can figure out the optimal partition point with the minimum latency for a given ith exit point.

Since the model partitioning does not affect the inference accuracy, we can sequentially try different DNN exit points according to the ordering of decreasing accuracy. Since a longer DNN model existing point ith generally possesses a higher accuracy ACC_i, p, we will select a longer model exit point earlier to find the model with the maximum accuracy meanwhile satisfying the latency requirement. As the regression models for layer-wise latency prediction have been trained in advance, Algorithm 7.6 mainly involves linear search operations and can be completed very quickly (no more than 1 ms in our experiment).

Algorithm 7.6 is guaranteed to find the optimal exit point and partition point. Intuitively, Algorithm 7.6 rolls the search first by exit point, and then exhaustively checks all potential partition points. The search of model exit point iterates from the longest one to the shortest one, so that the accuracy of the current selection is maximized. The search partition point is in a brute force manner, ensuring that the latency is minimized. Therefore, Algorithm 7.6 finds the optimal pair of the points. Under such two-phase search process, the complexity of Algorithm 7.6 is $O(MN)$, where M is the number of available exit points and N is the number of model layers.

Algorithm 7.6 Runtime Optimizer for Static Environment

Input:

M: the number of exit points in the DNN model

$\{N_i | i = 1, \cdots, M\}$: the number of layers in the branch of exit point i

$\{L_j | j = 1, \cdots, N_i\}$: the layers in the branch of exit point i

$\{D_j | j = 1, \cdots, N_i\}$: layer-wise output data size in the branch of exit point i

$f(L_j)$: the prediction model that returns the jth layer's latency

B: current available bandwidth

L: latency requirement

Output:

Selection of exit point and partition point

1: **Procedure**
2: **for** $i = M, \cdots, 1$ **do**
3:      Select the branch of ith exit point
4:      **for** $j = 1, \cdots, N_i$ **do**
5:          $ES_j \leftarrow f_{edge}(L_j)$
6:          $ED_j \leftarrow f_{device}(L_j)$
7:          $T_{i,p} = \text{argmin}_{p=1,\ldots,N_l}(\sum_{j=1}^{p-1} ES_j + \sum_{k=p}^{N_i} ED_j + D_{p-1}/B)$
8:          **if** $T_{i,p} \leq L$ **then**
9:              **return** Exit point i and partition point p
10: **return** NULL   ▷ cannot meet the latency requirement

7.3.3   EDGENT FOR DYNAMIC ENVIRONMENTS

The key idea of Edgent for the dynamic environment is to exploit the historical bandwidth traces and employ *Configuration Map Constructor* to generate the optimal co-inference plans for versatile bandwidth states in advance. Specifically, as shown in Figure 7.6, under the dynamic environment Edgnet generates the dynamic configurations (i.e., a configuration map that records the optimal plans for different bandwidth states) at the offline stage, which enables Edgnet to pick up the optimal partition plan according to the configuration map in run-time when at the online stage.

At the **offline configuration stage**, the dynamic configurator performs following tasks: (1) sketch the bandwidth states (noted as $s1, s2, \cdots$) from the historical bandwidth traces and (2) use the bandwidth states, latency requirement, and the DNN model as the input to acquire the optimal exit point and partition point. The representation of the bandwidth states is motivated by the existing study of adaptive video streaming [Akhtar et al., 2018], where the throughput of TCP connection can be modeled as a piece-wise stationary process where the connection consists of multiple non-overlapping stationary segments. In the dynamic configurator, it defines

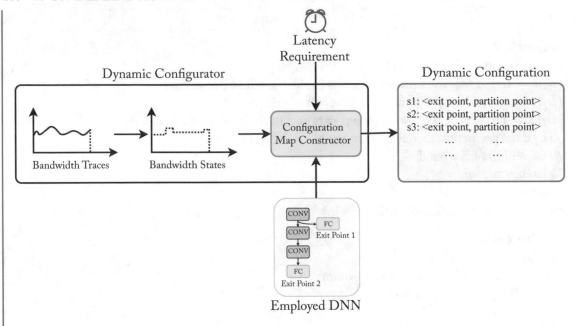

Figure 7.6: The dynamic configurator of Edgent.

a bandwidth state s as the mean of the throughput on the client side in a window segment of the underlying TCP connection. For all bandwidth states, we will call the optimization algorithm in *Configuration Map Constructor* to acquire the optimal co-inference plans and record these in a map/table as the dynamic configurations.

The optimization algorithm in *Configuration Map Constructor* is presented in Algorithm 7.7. The key idea of Algorithm 7.7 is to utilize the reward function to evaluate the selection of the exit point and partition point. Since our design goal is to maximize the inference accuracy while promising the application latency requirement, it is necessary to measure whether the searched co-inference strategy can meet the latency requirement and whether the inference accuracy has been maximized. Therefore, we define a reward to evaluate the performance of each search step as follows:

$$reward_{step} = \begin{cases} \exp{(ACC)} + throughput, & t_{step} \leq L, \\ 0, & else, \end{cases} \tag{7.2}$$

where t_{step} is the average execution latency in the current search step (i.e., the selected exit point and partition point in the current search step), which equals to $\frac{1}{throughput}$. The conditions of Equation (7.2) prioritizes that the latency requirement L should be satisfied, otherwise the reward will be set as 0 directly. Whenever the latency requirement is met, the reward of the current step will be calculated as $\exp{(ACC)} + throughput$, where ACC is the accuracy of current infer-

Algorithm 7.7 Configuration Map Construction

Input:
$\{s_i | i = 1, \ldots, N\}$: the bandwidth states
$\{C_j | j = 1, \ldots, M\}$: the co-inference strategy
$R(C_j)$: the reward of co-inference strategy C_j
Output:
Configuration Map

1: **Procedure**
2: **for** $i = 1, \ldots, N$ **do**
3:     Select the bandwidth state s_i
4:     $reward_{\max} = 0$, $C_{optimal} = 0$
5:     **for** $j = 1, \ldots, M$ **do**
6:         $reward_{c_j} \leftarrow R(C_j)$
7:         **if** $reward_{\max} \leq reward_{c_j}$ **then**
8:             $reward_{\max} = reward_{c_j}$, $C_{optimal} = C_j$
9:     Get the corresponding *exit point* and *partition point* of $C_{optimal}$
10:     Add $S_i :< exit\,point, partition\,point >$ to the Configuration Map
11: **return** Configuration Map

ence. If the latency requirement is satisfied, the search emphasizes on improving the accuracy. And when multiple options have the similar accuracy, the one with the higher throughput will be selected.[1] In Algorithm 7.7, s_i represents a bandwidth state extracted from the bandwidth traces and C_j is a co-inference strategy (i.e., a combination of exit point and partition point) indexed by j. $R(C_j)$ denotes the reward of the co-inference strategy C_j, which can be obtained by calculating Equation (7.2) according to the accuracy and the throughput of C_j.

Then at the **online tuning stage**, the *Runtime Optimizer* component selects the optimal co-inference plan according to the dynamic configuration and real-time bandwidth measurements. Algorithm 7.8 depicts the whole process in *Runtime Optimizer*. Note that Algorithm 7.8 calls the change point detection function $D(B_{1,\ldots,t})$ [Adams and MacKay, 2007] to detect the distributional state change of the underlying bandwidth dynamics. Particularly, when the sampling distribution of the bandwidth measurement has changed significantly, the change point detection function records a change point and logs a bandwidth state transition. Then with *find(state)* function, the *Runtime Optimizer* captures the corresponding co-inference strategy to the current bandwidth state (or the closest state) in the dynamic configuration and accordingly guides the collaborative inference process at the co-inference stage.

[1]In a dynamic environment, throughput is an important metric from the time average point of view.

Algorithm 7.8 Runtime Optimizer for Dynamic Environment

Input:

$\{B_{1,\cdots,t}\}$: the accumulated bandwidth measurements until the current moment t

$\{C_j | j = 1, \ldots, t\}$: the co-inference strategy

$\{s_i | i = 1, \ldots, t\}$: the bandwidth states

$D(B_{1,\ldots,t})$: the bandwidth state detection function that returns the current bandwidth state

$find(s)$: find the co-inference strategy corresponds to the given state s

Output:

Co-inference strategy

1: **Procedure**
2: $C_t = C_{t-1}$
3: $s_t = D(B_{i,\ldots,t})$
4: **if** $s_t \neq s_{t-1}$ **then**
5:     $C_t \leftarrow find(s_t)$
6: $s_{t-1} = s_t$
7: $C_{t-1} = C_t$
8: **return** C_t

7.4   PERFORMANCE EVALUATION

In this section, we present our implementation of Edgent and the evaluation results.

7.4.1   EXPERIMENTAL SETUP

We implement a prototype based on the Raspberry Pi and the desktop PC to demonstrate the feasibility and efficiency of Edgent. Equipped with a quad-core 3.40 GHz Intel processor and 8 GB RAM, a desktop PC is served as the edge server. Equipped with a quad-core 1.2 GHz ARM processor and 1 GB RAM, a Raspberry Pi 3 is used to act as a mobile device.

To set up a static bandwidth environment, we use the WonderShaper tool [Mulhollon, 2004] to control the available bandwidth. As for the dynamic bandwidth environment setting, we use the dataset of Belgium 4G/LTE bandwidth logs [van der Hooft et al., 2016] to emulate the online dynamic bandwidth environment. To generate the configuration map, we use the synthetic bandwidth traces provided by Oboe [Akhtar et al., 2018] to generate 428 bandwidth states range from 0–6 Mbps.

To obtain the branchy DNN, we employ BranchyNet [Teerapittayanon et al., 2016] framework and Chainer [Tokui et al., 2015] framework, which can well support multi-branchy DNN training. In our experiments we take the standard AlexNet [Krizhevsky et al., 2012] as the toy model and train the AlexNet model with five exit points for image classification over the

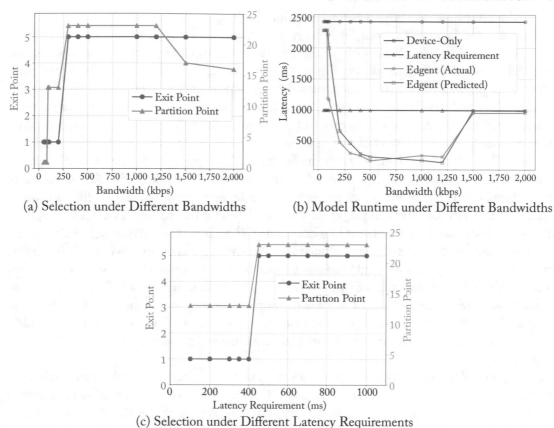

(a) Selection under Different Bandwidths   (b) Model Runtime under Different Bandwidths

(c) Selection under Different Latency Requirements

Figure 7.7: The results under different bandwidths and different latency requirements.

cifar-10 dataset [Krizhevsky et al., 2009]. As shown in Figure 7.3, the trained branchy AlexNet has five exit points, with each point corresponding to a branch of the standard AlexNet. From the longest branch to the shortest branch, the number of layers in each exit point is 22, 20, 19, 16, and 12, respectively.

7.4.2   EXPERIMENTS IN STATIC BANDWIDTH ENVIRONMENT

In the static configurator, the prediction model for layer-wise prediction is trained based on the independent variables presented in Table 7.1. The branchy AlexNet is deployed on both the edge server and the mobile device for performance evaluation. Specifically, due to the high-impact characteristics of the latency requirement and the available bandwidth during the optimization procedure, the performance of Edgent is measured under different pre-defined latency requirements and varying available bandwidth settings.

We first explore the impact of the bandwidth by fixing the latency requirement at 1000 ms and setting the bandwidth from 50 kbps to 1.5 Mbps. Figure 7.7a shows the optimal co-inference plan (i.e., the selection of partition point and exit point) generated by Edgent under various bandwidth settings. Shown in Figure 7.7a, as bandwidth increases, the optimal exit point becomes larger, indicating that a better network environment leads to a longer branch of the employed DNN and thus higher accuracy. Figure 7.7b shows the inference latency change trend where the latency first descends sharply and then climbs abruptly as the bandwidth increases. This fluctuation makes sense since the bottleneck of the system changes as the bandwidth becomes higher. When the bandwidth is smaller than 250 kbps, the optimization of Edgent is restricted by the poor communication condition and prefers to trade the high inference accuracy for low execution latency, for which the exit point is set as 3 rather than 5. As the bandwidth rises, the execution latency is no longer the bottleneck so that the exit point climbs to 5, implying a model with larger size and thus higher accuracy. There is another interesting result that the curve of predicted latency and the measured latency is nearly overlapping, which shows the effectiveness of our regression-based prediction. Next, we set the available bandwidth at 500 kbps and vary the latency requirement from 100–1000 ms for further exploration. Figure 7.7c shows the optimal partition point and exit point under different latency requirements. As illustrated in Figure 7.7c, both the optimal exit point and partition point climb higher as the latency requirement relaxes, which means that a later execution deadline will provide more room for accuracy improvement.

Figure 7.8 shows the model inference accuracy of different methods under different latency requirement settings (the bandwidth is fixed as 400 kbps). The accuracy is described in negative when the method cannot satisfy the latency requirement. As Figure 7.8 shows, given a tightly restrict latency requirement (e.g., 100 ms), all the four methods fail to meet the requirement, for which all the four squares lay below the standard line. However, as the latency requirement relaxes, Edgent works earlier than the other three methods (at the requirements of 200 ms and 300 ms) with the moderate loss of accuracy. When the latency requirement is set longer than 400 ms, all the methods except for device-only inference successfully finish execution in time.

7.4.3   EXPERIMENTS IN DYNAMIC BANDWIDTH ENVIRONMENT

For the configuration map generation, we use the bandwidth traces provided in Oboe [Akhtar et al., 2018]. Each bandwidth trace in the dataset consists of 49 pairs of data tuple about download chunks, including start time, end time, and the average bandwidth. We calculate the mean value of all the average bandwidth in the same bandwidth trace to represent the bandwidth state fluctuation, from which we obtain 428 bandwidth states range from 0–6 Mbps. According to Algorithm 7.7, through the exhaustive search, we figure out the optimal selection of each bandwidth state. The latency requirement in this experiment is also set to 1000 ms.

For online change point detection, we use the existing implementation [GitHub, 2013] and integrate it with the *Runtime Optimizer*. We use the Belgium 4G/LTE bandwidth logs

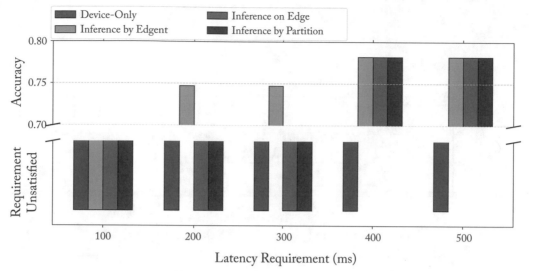

Figure 7.8: The accuracy comparison under various latency requirement.

dataset [van der Hooft et al., 2016] to perform online bandwidth measurement, which records the bandwidth traces that are measured on several types of transportation: on foot, bicycle, bus, train, or car. Additionally, since that most of the bandwidth logs are over 6 Mbps and in some cases even up to 95 Mbps, to adjust the edge computing scenario, in our experiment, we scale down the bandwidth of the logs and limit it in a range from 0–10 Mbps.

In this experiment Edgent runs in a dynamic bandwidth environment emulated by the adjusted Belgium 4G/LTE bandwidth logs. Figure 7.9a shows an example bandwidth trace on the dataset that is recorded on a running bus. Figure 7.9b shows the DNN model inference throughput results under the bandwidth environment showed in Figure 7.9a. The corresponding optimal selection of the exit point and partition point is presented in Figure 7.9c. Seen from Figure 7.9c, the optimal selection of model inference strategy varies with the bandwidth changes but the selected exit point stays at 5, which means that the network environment is good enough for Edgent to satisfy the latency requirement though the bandwidth fluctuates. In addition, since the exit point remains invariable, the inference accuracy also keeps stable. Dominated by our reward function design, the selection of partition points approximately follows the traces of the throughput result. The experimental results show the effectiveness of Edgent under the dynamic bandwidth environment.

We further compare the static configurator and the dynamic configurator under the dynamic bandwidth environment in Figure 7.10. We set the latency requirement as 1000 ms and record the throughput and the reward for the two configurators, based on which we calculate the Cumulative Distribution Function (CDF) to be. Seen from Figure 7.10a, under the same

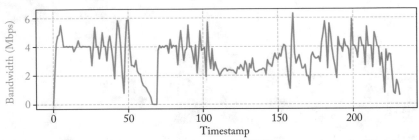

(a) An example bandwidth trace on Belgium 4G/LTE dataset (van der Hooft et al., 2016)

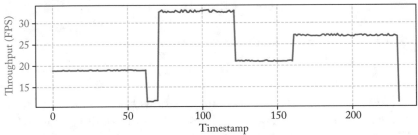

(b) The throughput of DNN model inference

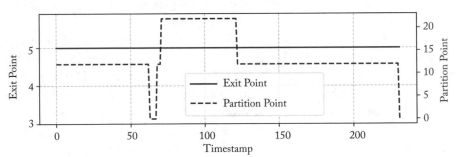

(c) The selection of exit point and partition point at each timestamp in the bandwidth traces

Figure 7.9: An example showing the decision-making ability of Edgent in a bandwidth traces recorded on a bus.

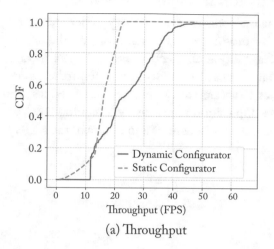

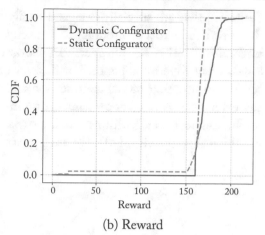

(a) Throughput   (b) Reward

Figure 7.10: The throughput and reward comparison between two configurations.

CDF level, Edgent with the dynamic configurator achieves higher throughput, demonstrating that under the dynamic network environment the dynamic configurator performs co-inference with higher efficiency. For example, set CDF as 0.6, the dynamic configurator makes 27 FPS throughput while the static configurator makes 17 FPS. In addition, the CDF curve of dynamic configurator rises with 11 FPS throughput while the static configurator begins with 1 FPS, which indicates that the dynamic configurator works more efficiently than the static configurator at the beginning.

Figure 7.10b presents the CDF results of reward. Similarly, under the same CDF level, the dynamic configurator acquires higher reward than the static configurator and the CDF curve of the dynamic configurator rises latter again. However, in Figure 7.10b the two curves are closer than those in Figure 7.10a, which means that the two configurators achieve nearly the same good performance from the perspective of reward. This is because the two configurators make similar choices in the selection of exit point (i.e., in most cases both of them select exit point 5 as part of the co-inference strategy). Therefore, the difference of the reward mainly comes from the throughput result. It demonstrates that the static configurator may perform as well as the dynamic configurator in some cases but the dynamic configurator is better in general under the dynamic network environment.

7.5   SUMMARY

In this chapter, we propose Edgent, an on-demand DNN co-inference framework with device-edge collaboration. Enabling low-latency edge intelligence, Edgent introduces two design knobs to optimize the DNN inference latency: DNN partitioning that enables device-edge collaboration, and DNN right-sizing that leverages early-exit mechanism. We introduce two configurators that are specially designed to figure out the collaboration strategy under static and dynamic bandwidth environments, respectively. Our prototype implementation and the experimental evaluation on Raspberry Pi shows the feasibility and effectiveness of Edgent toward low-latency edge intelligence. For future work, our proposed framework can be further combined with existing model compression techniques to accelerate DNN inference. Besides, we can extend our framework to support multi-device application scenarios by designing efficient resource allocation algorithms. We hope to stimulate more discussions and efforts in the society to fully realize the vision of edge intelligence.

CHAPTER 8

Applications, Marketplaces, and Future Directions of Edge Intelligence

Having presented a review of existing efforts (with bias toward our own research) on edge intelligence (which is still in its infancy stage), we next share our view of its applications, marketplaces, and future research directions to conclude this book. Edge intelligence is an emerging interdisciplinary field with many open problems and yet tremendous opportunities, and our study in this monograph touches only the tip of iceberg.

8.1 APPLICATIONS OF EDGE INTELLIGENCE

In this section, we discuss applications enabled by edge intelligence, together with their real-world use cases.

8.1.1 VIDEO ANALYTICS

Surveillance cameras are ubiquitous, and many cities and organizations are deploying unprecedented amounts of cameras for a wide variety of purposes, including public safety, security surveillance, traffic control, and planning and factory floor monitoring. The shipments of surveillance cameras are expected to grow at more than 20% compound annual growth rate from 2017–2024 [IHS Markit, 2018]. Driven by the recent advances in computer vision and deep learning, a continuous stream of increasingly accurate DNN models for object recognition (e.g., ResNet152 and VGG-16) have been proposed, enabling video analytics with even greater potential. However, due to the huge and ever-growing size of video data, video analytics only on the cloud is no longer practical, due to the concerns on both performance and cost. Specifically, transferring the large volume of video data via the already overwhelmed WAN to the cloud incurs prohibitive bandwidth cost and unacceptable transmission latency.

The recent years have witnessed a shift of moving video analytics from the cloud to the edge [Ananthanarayanan et al., 2017, Hung et al., 2018, Jiang et al., 2018a,b, Quan et al., 2018, Yi et al., 2017, Zhang et al., 2018c]. Apart from reduced monetary cost and transmission latency, edge computing further empowers video analytics with high responsiveness, location awareness, and context awareness. These virtues are quite meaningful to mission-critical tasks

Table 8.1: Applications of edge intelligence

EI Application	AI Basis	Commercial Products	Category
Video Analytics	CV	SenseTime SenseDLC SDK	Software
		HIKVISION iDS Smart Camera	Device
		Dell EMC Video Surveillance Solution	Platform
		Huawei Video Surveillance Solution	Platform
		Honeywell Video Surveillance Solution	Platform
Cognitive Assistance	CV, NLP, SR	Google Lens	Device
		Amazon Alexa	Software
		Microsoft Cortana	Software
		Apple Siri	Software
IIoT	CV, Robotics	Intel Industrial Automation	Platform
		IBM PMO	Platform
		Alibaba Cloud ET Industrial Brain	Platform
		Simens MindSphere	Platform
Smart Home	CV, NLP, SR, Robotics	Intel Smart Home Development Acceleration Platform	Platform
		Huawei Smart Home Solution	Platform
		Google Home Hub	Software
		Microsoft Home Hub	Software
Precision Agriculture	CV, Robotics	DJI Agriculture Drone	Device
		Intel Precision Agriculture Kit	Software
		KAA Smart Farming Solution	Software
Smart Retail	CV, NLP, SR, Robotics	Intel Smart Retail Solution	Platform
		Amazon Go	Infrastructure
		JD.com Unmanned Store	Infrastructure

in many fields. Taking the "Amber Alert" for child abduction in public safely for example, by vehicle recognition, vehicle license plate recognition and face recognition using various connected cameras at the interested regions, edge computing can identify suspects and suspicious vehicles or persons with the quickest responses [Zhang et al., 2018c]. Due to the great potential of edge-based live video analytics, it is recognized as the "killer application" of edge computing by Microsoft [Ananthanarayanan et al., 2017]. More excitingly, realistic deployments of edge-based video analytics have been conducted in cities like Bellevue, WA, Washington, DC, Cambridge in UK, and Beijing and Hangzhou in China.

Use Cases: In partnership with the City of Bellevue, WA and City of Washington, DC, Microsoft has developed traffic control solutions for the two cities, based on traffic video analytics running on top of edge clusters. By understanding volumes of cars, buses, trucks, motorcycles, bicycles, walkers, and near-miss collisions, they proposed solution predicts where future crashes are likely to occur. Actuation systems could then take corrective action (e.g., traffic light duration control) to prevent the collisions.

8.1.2   COGNITIVE ASSISTANCE

Digital experiences enhanced by cognitive assistance are considered to be among the most promising trends in consumer electronics, as demonstrated by Apple Siri, Google Lens, Microsoft Cortona, and Amazon Alexa. Cognitive assistance brings AI technologies such as speech recognition, emotion detection, computer vision, and natural language processing within the inner loop of human cognition and interaction. By analyzing sensor data from a wearable device or a smartphone in a real-time manner, cognitive assistance performs a task required by the users, or returns just-in-time task-specific guidance to the user. For example, Apple Siri can automatically make a phone call based on the voice command of the user. Microsoft Cortona can remind and direct the user to drive to the airport in advance, based on the email of the flight ticket order. With the on-device camera, Google Lens enables visual search by recognizing the interested objects (e.g., a beautiful clothe in the behind the show window) in the camera view and rendering related information.

With cognitive assistance, many tasks that only humans used to be able to do can be undertaken by wearable or mobile devices. This greatly enriches people's life style, improves human production efficiency, and brings unlimited opportunities for business. A distinctive feature of cognitive assistance is that it is highly interactive and thus requires low latency. However, since cognitive assistance applications typically performs both computational- and data-intensive AI tasks (e.g., vision and audio processing), neither local execution nor cloud offloading is an ideal solution for implementation. Specifically, running on resource- and power-constrained mobile or wearable devices prolongs the data processing time while running on the cloud prolongs the data transmission time. Currently, an emerging class of interactive cognitive assistance applications are poised to edge computing infrastructures, becoming the key demonstrators of edge computing [Chen et al., 2015c, 2017b, Guo et al., 2018, Ha et al., 2014].

8.1.3   INDUSTRIAL INTERNET OF THINGS (IIoT)

Industry 4.0, also known as smart manufacturing or smart industry, is a vogue word representing the current trend of revolution from traditional manufacturing to intelligent manufacturing. An important element of Industry 4.0 is the Industrial Internet of Things (IIoT) that interconnects sensors, instruments, and other devices together with industrial production applications. By empowering IIoT with AI and big data capabilities, we can make production systems high accurate and intelligent, which directly leads to increased efficiency and productivity in assembly/product lines, as well as decreased maintenance and operation expenditures.

For example, for machine diagnosis, we can use computer vision techniques to recognize the shape change and crack that may lead to the shutdown of machines (e.g., mechanical arms and cutting tools) [Shao et al., 2016]. By doing so, we can diagnosis and maintain machines in a proactive manner, which can greatly reduce the downtime of machines, and thus improve the production efficiency. Another example is the manufacture inspection for quality assurance of products [Li et al., 2018d, Luckow et al., 2017]. Traditionally, this was performed in an artificial manner, i.e., by hiring inspectors. Today, with edge intelligence, this can be done by using cameras to detect the defects or products passed the assembly line, with greatly improved cost-efficiency.

Use Cases: IBM Predictive Maintenance and Optimization (PMO) is an IIoT solution based on Watson IoT Platform Edge. It enables manufacturing organizations to apply edge intelligence to improve quality and maintenance management with minimal cost. Currently, this solution has been applied by Shenzhen China Star Optoelectronics Technology Co., Ltd. to accelerate and automate quality inspection of the produced LCD screens. The technique significantly boosts production quality and throughput, while cutting management and maintenance cost.

8.1.4   SMART HOME

Smart home is also known as home automation, a smart home system controlling lighting, climate, security, entertainment systems, and appliances in an intelligent manner. Nowadays, several smart home appliances that are built on the basis of IoT concept are already commercially available, as exemplified by smart light, internet TV, smart washer, and robot vacuum, etc. However, most of current practices on smart home just equip the electrical devices with wireless communication modules, connect them to the cloud, and enable them with remote control capabilities. Clearly, this is insufficient to fully unleash the potential of smart home.

With edge computing in a home environment, smart devices can be connected to the edge computing node which can be a gateway or an access point. By aggregating the data generated by various appliance at the edge computing node and analyzing them in a coordinated manner, more sophisticated services can be provided [Jie et al., 2017a, Xu et al., 2018]. For example, the electricity bill can be reduced by intelligently scheduling the workload of multiple appliance to exploit the time-varying electricity price. This goal can be achieved by mining the usage pattern of various appliance from the historical data, and then temporally scheduling the workload in a predictive manner to reduce the electricity expenditure, without harming the user experience [Zhang et al., 2017b]. Besides, with edge intelligence, cheap wireless sensors and controllers can be deployed to more corners of the house, making fine-grained control of the smart home possible [Jie et al., 2017b]. For instance, with more temperature sensors deployed, we can set the temperature of air conditioners in different rooms in a fine-grained manner to improve the comfort level, based on the detected activities. Finally, for low-latency appliance such as vision or voice recognition based smart lock, edge intelligence brings fast response to them.

Use Cases: Intel Smart Home Development Acceleration Platform is a software development kit initially available on Intel Quark™ processors, which is designed to connect devices in the home. Empowered by AI capabilities such as speech, vision, and cognition, the kit transforms homes from being merely connected to becoming smart, and brings new ways for consumers to interact with home appliances around them. In collaboration with the Amazon Alexa Voice Service, this technique has been integrated into Alexa (a virtual assistant developed by Amazon), bringing the convenience of voice control to any connected home appliance.

8.1.5   PRECISION AGRICULTURE

Boosted by the growing population and upward social mobility, the food demand is expected to double by 2050. Clearly, if agriculture is to continue to feed the world, it needs to become more productive. Fortunately, that is already beginning to happen. In addition to bio-techniques, data and intelligence techniques are bringing new ways to farmers to increase yields, improve quality and reduce crop damage. These data- and AI-drive techniques are known as precision agriculture. As predicted by the International Food Policy Research Institute, precision agriculture can help us to meet the doubled food demand via increasing farm productivity by as much as 67% by 2050 and cutting down agricultural losses [Vasisht et al., 2017].

Today, edge intelligence has been applied to various major aspects of farming to promote precision agriculture. For example, in irrigation water management, edge intelligence can better assess need and availability of soil water level for crop cultivation. Using the time-series of meteorological data sensed by underground sensors, the future soil moisture level can be predicted using AI methods such as LSTM and DNN, such that proactive and precise irrigation can be performed in advance [Mohapatra et al., 2018]. Another example is the prevention of diseases and pests. Using agricultural drones to scanning crops with visible and Infrared (IR) light, a precise crop image can be constructed during the flying. Then, by analyzing the image with computer vision techniques, we can detect which plants may be infected by bacteria or fungus, helping to prevent disease from spreading to other crops. In addition, edge intelligence can be also applied to crop detection and classification [Bosilj et al., 2018], detection of fruit stages (raw or ripe) [Sa et al., 2016], automated harvesting, etc.

Use Cases: Baidu Intelligent Edge (BIE) is an open-source computing platform aiming at environmental detection tasks for cars, drones, and other moving systems. As a major component of BIE, the object recognition module BIE-AI-Board is compact enough to be embedded into cameras, drones, robots, and other edge devices. Early partnership with an agricultural drone company has applied this technique to intelligent crop damage detection, the results is quite inspiring—reducing the usage of pesticide by up to 50%.

8.1.6 SMART RETAIL

In today's fiercely competitive marketplace of retail, traditional brick-and-mortar stores need to find new opportunities to improve efficiency, reduce cost and enhance shopping experience. A recent shift toward these goals is smart retaining, which describes a set of smart technologies that are designed to give the consumer a greater, faster, safer, and smarter experience when shopping. To enable smart retail, both IoT techniques that track information of products and users, and AI approaches that analyze the acquired user and product data, are very essential.

Edge intelligence is able to improve all the aspects of retailing, ranging from marketing, payment, inventory, and cost management to layout optimization, etc. Here we use a simple and realistic example to show how edge intelligence can transform the retailing industry. Specifically, when Alice is approaching the geographical reach of a smart shop, proximity-activated promotion would push notifications to her smartphones. If Alice then passes by the smart shop, augmented reality would capture her and deliver the personalized experience of recommended products to her. Next, if Alice really stepped into the shop, a shopping robot may ask about her interested products and direct her to corresponding shelf. Meanwhile, the cameras in the store detect which item she takes or places back the shelf and compute the payment. Then, after Alice pays online and leaves the shop without being checked out by a cashier, the central controller may turn down the lighting and HAVC systems to reduce the energy cost. Finally, by further mining the shopping data, we can predict the popularities of various products, which will help to improve the layout of products and inventory efficiency.

Use Cases: As a basic realization of smart retail, unmanned store is taking off in China's retail market. In particular, the internet giants Alibaba, Tecent and JD.com opened unmanned stores since 2017. Powered by smart shelves, sensors, cameras, digital signages, and cashier-less checkout counters, these stores offer an unprecedented level of convenience for customers—they can simply pick up whatever they want and walk straight out of the store without getting slowed down by queues or payments.

Beyond the above emerging applications, we should note that *edge intelligence is increasingly boosting many other applications in various vertical areas, such as autonomous driving [Huang et al., 2017b, Zhang et al., 2018b], smart healthcare [Chang et al., 2018], smart city [Hossain et al., 2018], smart grid [He et al., 2017], intelligent robots [Chen et al., 2018c], sport and entertainment [Gowda et al., 2017], etc.* Due to the space limit, we do not introduce the applications of edge intelligence in these areas one by one, and interested readers are referred to the references therein.

8.2   MARKETPLACE OF EDGE INTELLIGENCE

In this section, we first overview the marketplace of Edge Intelligence with concentration on several supporting fields therein. Then we introduce some existing commercial hardware and platforms for EI and characterize them with distinctive features.

8.2.1   MARKET OVERVIEW AND LANDSCAPE

Since EI is the nexus between AI and edge computing, any field that is relevant to AI or edge computing can have a potential impact on the market of EI. However, instead of attempting to grasp the whole market in a few words, we narrow down our view and focus only on fields that are the most critical for EI, namely hardware, software, and platforms on which EI relies. Other market fields like infrastructure, connectivity, traffic routing and delivery, virtualization and containerization, etc. should be orthogonal to EI, and we refer readers to a community-driven state of the edge computing landscape for more details [GitHub, 2018].

Hardware for EI

The most straightforward way to launch EI is to run AI algorithms directly on edge devices. However, edge devices are typically resource-constrained while the latest advanced AI techniques such as DNN models are resource-hungry owing to their complex structures and massive numbers of parameters. Though the workload of AI algorithms poses an obstacle to realizing EI, some startups as well as big companies still see chances in it and have brought AI chips with variant architectures for the edge to the market. Designed for meeting the extreme requirements of heterogeneous edge devices, AI chips for EI are typically purposely optimized for inferencing tasks in a power-efficient manner. The recent years have witnessed more and more startups and big companies stepping into the field of chips dedicated to running AI tasks on resource-constrained devices. Examples include startups like Cambricon and Horizon Robotics, as well as big companies like Google and Intel, who introduced Edge TPU and Movidius Myriad VPU, respectively. A detailed discussion on AI chips for EI is deferred to Section 8.2.3.

Software for EI

Different from the hardware approach that tends to extend the capability of edge devices, there are also efforts on enabling EI in a perspective of software, which can be roughly grouped into two types, namely efforts on reducing the sizes of AI models and on handling the heterogeneity of the architectures of edge devices.

Reducing the sizes of AI models indicates adapting AI models for edge devices with various means (as exemplified by parameter binarization and branchy architecture), so that the models can better fit the capabilities of edge devices without compromising on performance. Companies focusing on reducing AI model sizes include XNOR.AI, etc.

Owing to the natural heterogeneity of edge devices, deploying the same AI models onto edge devices can be problematic and requires additional labors. To ease the deployment of AI models on edge devices, an abstraction layer such as containerization for edge runtime is needed so that AI models can be deployed in a hardware-agnostic manner. Projects or companies focusing on the heterogeneity of edge devices include balenaOS, etc.

Platforms for EI

Besides running AI tasks purely on edge devices with the help of energy-efficient yet powerful AI chips and models tailored for edge devices, there are alternatives for achieving EI which follow the computing paradigm of edge computing and edge-cloud computing, i.e., delegating the computation-intensive AI tasks to nearby edge servers and further offloading tasks to the powerful remote cloud when appropriate.

There are Edge PaaS companies which provide AI-powered edge platforms, so-called platforms for EI, such as Google Cloud IoT Edge and FogHorn. A detailed discussion on platforms for EI is shown in Section 8.2.2.

8.2.2   EMERGING COMMERCIAL PLATFORMS FOR EI

In this section, we first introduce a list of features for EI platforms, and then give some examples and classify existing commercial platforms for EI based on these features.

Features for EI Platforms

We propose a list of features for EI platforms and if a platform meets all of these features, we regard it as a general EI platform.

Features inherited from IoT platforms. Since IoT has become a top-level scenario and is no doubt one of the biggest data generators that can benefit much from both edge computing and AI techniques, we argue that a general EI platform should first be capable of managing a massive number of IoT devices. The features required by an IoT platform usually include supports for real-time responding at the device-side; offline or intermittent-connectivity; multi-layered security of device data; communication within local devices and communication between local devices and the cloud; cloud-based data aggregation, management, and analytics; and device management and maintenance at scale.

Flexible cloud-edge-device synergy. This is the feature required by edge computing where (partial) compute happens at or in proximity to the sources of data, which is beneficial to enabling real-time responding and filtering out unnecessary data transmission. More specifically, the support for local compute and edge-cloud synergy should be flexible and general since edge devices are naturally heterogeneous, which is to say, are of different architectures, as well as variant computation capacities, storage sizes, power supplies, etc.

Explicit support for AI techniques. This feature is posed by EI, of course, and the intelligence can be achieved at the edge side, at the cloud side, or both.

In the following, we give an example of general commercial EI platforms, an example of EI platforms dedicated for some specific applications, and at last, a table of existing EI platforms.

An Example of General EI Platforms

We shortly introduce Google Cloud IoT Edge, which meets all the features proposed for EI and is so-called a general platform. Logically,[1] the Google Cloud IoT Edge platform can be seen as a combination of three layers, namely the sensors layer, the edge layer, and the cloud layer. The sensors layer is responsible for sensing the environment while the edge layer gathers data from sensors and performs AI-powered analytics to enable intelligent decision made locally in no time. The data collected by edge devices are further uploaded to the remote cloud so that the ML models can be trained incrementally and the configuration of the system can be tuned with new data. Besides, the platform has special support for Edge TPU, an AI chip for EI provided by Google, in its edge runtime and management of a massive number of IoT devices is also supported. An illustration of the architecture of Google Cloud IoT Edge platform is shown in Figure 8.1.

An Example of Dedicated EI Platforms

We take Huawei's video surveillance platform, which enables intelligent video surveillance via edge-cloud synergy, as an example of dedicated EI platforms. At the edge side, Huawei's AI-enabled software-defined cameras, the M/X series, use deep learning based computer vision technologies to analyze a large amount of video in real time. The analytics helps extract valuable information from the video and alleviates the pressure of both the network and the cloud. Therefore, the cloud video nodes can handle more cameras and focus on higher-level applications such as video searching among multiple sites.

Commercial EI Platforms

A list of example commercial EI platforms are shown in Table 8.2.

8.2.3   EMERGING COMMERCIAL DEVICES AND HARDWARE FOR EI

In this section, we give a hierarchy of AI chips characterized by their target scenarios, and then divide some commercial AI chips into these defined levels accordingly.

---

[1]Sensors and edge devices need not be separated physically.

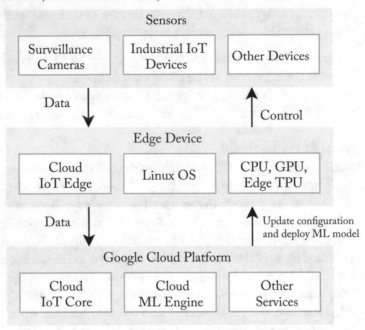

Figure 8.1: The architecture of Google Cloud IoT Edge platform.

The Hierarchy of AI Chips

Targeting different use-case scenarios, AI chips in the market can be roughly divided into three categories, namely wearable, mobile-device, and edge-infrastructure. Their main differences lie on the compute capability and the power budget.

AI chips for wearable such as wristbands typically provide compute capability ranging from 10MOPS–100GOPS with the power budget below 10 mW. They are usually designed only for inference tasks and can serve models with sizes ranging from 10–100 KB.

AI chips for mobile-device such as IoT devices, smartphones, and mobile VR/AR devices typically deliver compute capability ranging from 1–10TOPS with the power budget around 1–10 W. AI chips for edge-devices can serve models with sizes about 10 MB.

AI chips for edge-infrastructure typically deliver compute capability starting from 10TOPS with the power budget exceeding 10 W. AI chips for edge-infrastructure often have full support for variant inference tasks offloaded from resource-constrained wearable or mobile devices.

Table 8.2: Commercial platforms for EI

Type	Use-Case	EI Platforms
General	Industrial IoT Intelligent V2I Smart City Unmanned Retail	Aliyun Link Edge AWS IoT Greengrass Baetyl with Baidu Intelligence FogHorn Lightning Edge AI GE Predix Google Cloud IoT Edge Huawei Intelligent EdgeFabric IBM Edge Application Manager Microsoft Azure IoT Edge Smartiply Fog Platform
Dedicated	Facial Recognition System	SenseTime SenseFace Face Recognition Surveillance Platform
	Video Surveillance	Huawei Video Surveillance Platform

Commercial AI Chips in Hierarchy

Table 8.3 summarizes a list of representative commercial AI chips and SoCs, which are divided into different levels according to the hierarchy and further characterized by detailed use-case scenarios.

8.3 FUTURE DIRECTIONS ON EDGE INTELLIGENCE

Based on the applications and marketplaces discussed above, we now present our view of the open challenges and future directions of edge intelligence.

8.3.1 BUILDING A THEORETIC FOUNDATION OF EDGE INTELLIGENCE (EI)

As noted above, pushing the AI frontier to the edge ecosystem that resides at the last mile of the Internet, is highly non-trivial, and there is no "one-level-fits-all" edge AI solution for all applications. The past few years have witnessed significant advances in EI, thanks to interdisciplinary effort in computer science, statistics, optimization, and other disciplines. Despite ever increasing research effort on EI, a fundamental understanding of EI continues to elude us. Driven by many emerging real-time EI applications (including autonomous driving, smart robots, safety-critical health applications, and augmented/virtual reality), the need for building a theoretic foundation for EI becomes even more pressing.

Table 8.3: Commercial AI chips for EI

Hierarchy	Use-Case	AI Chips
Wearable	ECG Analysis Fall Detection Intelligent Personal Assistant	Huami MHS001 Huawei Ascend Nano & Tiny Intel Movidius Myriad X VPU
Mobile-device	Autonomous Driving Machine Vision Robotics Voice Recognition	Cambricon-1H/1M Google Edge TPU Horizon Robotics Journey/Sunrise Huawei Ascend Lite/Mini Intel Movidius Myriad VPU Intel Mobileye EyeQ NVIDIA Jetson Modules Xilinx Zynq SoC and MPSoC Edge Cards
Edge-infrastructure	Model Tuning Offloaded Inference Tasks	Baidu Kunlun Huawei Ascend Multi-Mini/Max Intel Nervana NVIDIA Tesla P4

Resource-Friendly EI Model Design

Many existing AI neural network models such as CNN and LSTM were originally designed for applications such as computer vision and natural language processing. Most of deep learning based AI models are highly resource-intensive, which means that powerful computing capability supported by abundant hardware resources (e.g., GPU, FPGA, TPU) is an important boost the performance of these AI models.

Building on these advances, we advocate a resource-aware EI model design. Instead of utilizing the existing resource-intensive AI models, we can leverage the AutoML idea [He et al., 2018] and the Neural Architecture Search (NAS) techniques [Zoph and Le, 2016] to devise resource-efficient EI models tailored to the hardware resource constraints of the underlying edge devices and servers. For example, methods such as reinforcement learning, genetic algorithm, and Bayesian optimization can be adopted to efficiently search over the AI model design parameter space (i.e., AI model components and their connections) by taking into account the impact of hardware resource (e.g., CPU, memory) constraints on the performance metrics such as execution latency and energy overhead.

For deep learning applications running on the cloud, container technology has become the de-facto standard for application deployment. For example, when training a large DNN model in a distributed manner, the dataset or model can be partitioned into multiple parts, each to be trained locally on a container, and results on different containers are aggregated to update the global model. The advantages of container technology are multifold: enabling efficient server resource sharing among multiple applications with little performance interference, improved flexibility and cost efficiency, etc. Due to these advantages, container has been widely recognized as the best option for resource allocation and application deployment in edge computing.

However, for edge intelligence, using container technology is facing new challenges. Due to the high capital expenditure of GPU and high-end CPU chips, it is desirable that each GPU card can be efficiently shared by multiple IoT devices. e.g., using a GPU card to support the video analytics for video feeds from multiple (e.g., 8 or 16) cameras. The use case is quite different from that in cloud computing. Since workload in the cloud is typically computational demand, the current practice of containerization is to allocate one or more CPU/GPU cards for each container. While in edge computing, since the workload (e.g., DNN inference) is often lightweight, it is highly desired that each CPU/GPU card can be shared by multiple containers. In this case, the resource overhead of containers themselves is non-neglectable, impeding efficient resource sharing among multiple containers. Specifically, as per our private conversations with industrial partners, for a video analytics application, the resource of a GPU card becomes almost saturated if running three containers to process high-resolution video feeds from three cameras. Therefore, toward efficient edge resource sharing, more lightweight containerization technology can be an appealing research direction for edge intelligence.

High Dimensional Model Configuration and Optimization for Edge Intelligence
As we have discussed in previous chapters, to optimize the performance of edge intelligence, various optimization techniques can be applied and customized to different workloads and hardware platforms. Clearly, to maximize the overall performance of the edge intelligence system, various optimization techniques should work in a cooperative manner to provide greater flexibility and optimization space. When jointly applying multiple optimization techniques, we face a high-dimensional configuration problem that is required to determine a large number of performance-critical configuration parameters. Taking video analytics for example, the high-dimensional configuration parameters can include video frame rate, resolution, model selection and model early-exit, etc.

High dimensional configurations put great burdens on users and developers, who are now responsible for setting performance-critical configurations to ensure the system is performant and available. Unfortunately, this burden is more than that most end users can handle, due to the following challenges: (1) high-dimensional configurations represent tradeoffs, e.g., for video analytics, a lower video resolution decreases the resource cost and latency, but at the cost of reduced recognition accuracy; (2) to manage those tradeoffs among multiple performance metrics, deep knowledge of the underlying hardware, workload, and relationships between performance metrics and configurations are required. However, due to the complex interactions between performance, workload, configurations and platforms, these relationships are often too complex to be clearly explained in the documentation or mathematically formulated. Even with clear documentation, due to the dynamical environment, those relationships can be time-varying, i.e., changing quickly, making it is challenging for programmers to maintain a proper setting. In many cases, there is simply no satisfactory static setting for high-dimensional configurations; and (3) due to the combinatorial nature, high-dimensional configuration problem involves a huge search space of parameters. How to search a near-optimal solution with a low computation overhead is highly desirable. Currently, high-dimensional configuration problems have been studied in the context of datacenter resources management, machine learning and control-theoretic based solutions are proven to be effective. While these solutions provide insights for the problem in edge intelligence, we should note that the later typically poses more stringent performance requirement with lower resource supply. Therefore, how to adapt existing solutions to edge intelligence still deserves further investigation.

In what follows, we use DNN design as an example to elaborate on the trade-off among various DNN performance metrics. For an edge intelligence application with a specific mission, there is usually a series of DNN model candidates that are capable of finishing the task. However, it is difficult for software developers to choose an appropriate DNN model for the EI application because the standard performance indicators such as top-k accuracy or mean average precision fail to reflect the runtime performance of DNN model inference on edge devices. For instance, during the EI application deployment phase, beside accuracy, inference speed and resource usage are also critical metrics. We need to explore the trade-offs between these metrics and identify the factors that affect them. In the effort [Huang et al., 2017a], for the object recognition application, the authors investigate the influence of the main factors, e.g., number of proposals, input image size and the selection of feature extractor, on inference speed and accuracy. Based on their experiment results, a new combination of these factors is found to outperform the state-of-the-art method. Therefore, it is necessary to explore the trade-offs between different metrics, helping to improve the efficiency of deploying EI applications.

Resource-Friendly Machine Learning Approaches for Edge Intelligence

Due to the unique challenges in edge networks, model training and inference therein need to be cognizant of the communication, computation, and architectural features of the network edge, and can take different forms, such as centralized, distributed, or hybrid. Of particular interest are distributed learning approaches, including federated, collaborative, and distributed learning (e.g., multi-agent learning), for techniques that range from DNNs, reinforcement learning, imitation learning, graph neural networks, and GANs—to name a few.

To get a more concrete sense, in what follows we use wireless edge learning to illuminate how reinforcement learning can be carried out in wireless networks. Traditional network approaches are usually based on some assumptions to develop policies for wireless network management. ML approaches would be required to adapt, in real time, to the fast time-varying statistics of wireless channels and the dynamics of diverse network traffic. Further, the data samples available for adaptation are often limited, calling for a synergy of data-driven ML methods with classical model-based approaches. RL is one potential alternative approach that imposes fewer assumptions and has a strong adaptive capability. For example, assuming that we already have some prior models, by using RL, one can learn the time varying network models in real time and adapt the policies accordingly, which could converge to a desirable solution for a particular network. In addition, by using reinforcement learning, we can create a closed-loop between a learning policy and the network environment, which is more robust and adaptive to dynamic environments. RL could work well because the network management problem is often more "structured" than in other fields where RL has not succeeded. Moreover, by integrating RL with federated learning to form a new paradigm of federated RL, we can achieve efficient global knowledge sharing from multiple edge nodes to enhance the performance further. By making use of structural domain knowledge, machine learning approaches move from conventional black-box to gray/white-box approaches, offering promises in addressing the aforementioned challenges.

In Table 8.4, we outline the highlights of some popular machine learning approaches for model training/inference toward achieving edge intelligence, including the ones reviewed in Chapter 2. In a nutshell, there is an urgent need to develop a theoretical foundation to understand the above ML techniques and beyond, so we can quantify the model accuracy in edge environments.

8.3.2   LEARNING-DRIVEN NETWORKING, SECURITY, AND PRIVACY TECHNIQUES FOR EI

For EI, computation-intensive AI-based applications are typically running on the distributed edge computing environment. As a result, advanced networking solutions with computation awareness are highly desirable, such that the computation results and data can be efficiently shared across different edge nodes.

For the future 5G networks, the Ultra-Reliable Low-Latency Communication (URLLC) has been defined for mission-critical application scenarios that demand low delay and high reliability. Therefore, it will be promising to integrate the 5G URLLC capability with edge computing to provide Ultra-Reliable Low-Latency EI (URLL-EI) services. Also, advanced techniques such as software-defined network and network function virtualization will be adopted in 5G. These techniques will enable flexible control over the network resources for supporting on-demand interconnections across different edge nodes for computation-intensive AI applications.

On the other hand, autonomous networking mechanism design is important to achieve efficient EI service provisioning under dynamic heterogeneous network coexistence (e.g., LTE/5G/WiFi/LoRa), allowing newly added edge nodes and devices to self-configure in the plug and play manner. Also, the computation-aware communication techniques are starting to draw attention, such as gradient coding [Tandon et al., 2017] to mitigate straggler effect in dis-

Table 8.4: Resource-friendly ML approaches for edge intelligence

ML Methods	Highlights
Federated Learning	• Leave training data distributed on end devices • Train shared model on server by aggregating locally computed updates • Preserve privacy
Aggregation Frequency Control	• Determine trade-off between local update and global parameter aggregation under given resource budget • Intelligent communication control
Gradient Compression	• Quantizing each element of gradient vectors to a finite-bit low precision value • Gradient sparsification by transmitting only some values of the gradient vectors
DNN Splitting	• Select a splitting point to reduce latency as much as possible • Preserve privacy
Knowledge Transfer Learning	• First train a base network (teacher network) on a base dataset and task and then transfer the learned features to a second target network (student network) to be trained on a target dataset and task • The transition from generality to specificity
Gossip Training	• Random gossip communication among devices • Full asynchronization and total decentralization • Preserve privacy
Edge-cloud Collaboration	• Utilize the cloud to tackle performance limitation of edge devices • Efficient communication process • Preserve privacy
Meta-Learning	• Optimize for the ability to learn how to learn by exploiting correlated tasks • Enable fast learning with small sample size
Reinforcement Learning	• Consist of policy, reward, state, action, environment model • Enable sequential decision making under in stochastic dynamic environment

tributed learning and over-the-air computation for distributed stochastic gradient descent [Zhu et al., 2018], which can be useful for edge AI model training acceleration.

The open nature of edge computing imposes that the decentralized trust is required such that the EI services provided by different entities are trustworthy [Li et al., 2018a]. Thus, lightweight and distributed security mechanism designs are critical to ensure user authentication and access control, model and data integrity, and mutual platform verification for EI. Also, it is important to study novel secure routing schemes and trust network topologies for EI service delivery when considering the coexistence of trusted edge nodes with malicious ones.

On the other hand, the end users and devices would generate a massive volume of data at the network edge, and these data can be privacy sensitive since they may contain users' location data, health or activities records, or manufacturing information, among many others. Subject to the privacy protection requirement, e.g., EU's General Data Protection Regulation (GDPR), directly sharing the original datasets among multiple edge nodes can have a high risk of privacy leakage. Thus, federated learning can be a feasible paradigm for privacy-friendly distributed data training such that the original datasets are kept in their generated devices/nodes and the edge AI model parameters are shared. To further enhance the data privacy, more and more research efforts are devoted to utilizing the tools of differential privacy, homomorphic encryption and secure multi-party computation for designing privacy-preserving AI model parameter sharing schemes [Du et al., 2018].

8.3.3   PROGRAMMING AND SOFTWARE PLATFORMS FOR EI

Currently, many companies around the world focus on the AI cloud computing service provisioning. Some leading companies are also starting to provide programming/software platforms to deliver edge computing services, such as Amazon's Greengrass, Microsoft's Azure IoT Edge and Google's Cloud IoT Edge. Nevertheless, currently, most of these platforms mainly serve as relays for connecting to the powerful cloud data centers.

As more and more AI-driven computation-intensive mobile and IoT applications are emerging, edge intelligence as a service (EIaaS) can become a pervasive paradigm and EI platforms with powerful edge AI functionalities will be developed and deployed. This is substantially different from machine learning as a service (MLaaS) provided by public clouds. Essentially, MLaaS belongs to cloud intelligence and it focuses on selecting the proper server configuration and machine learning framework to train model in the cloud in a cost-efficient manner. While in a sharp contrast, EIaaS concerns more about how to perform model training and inference in resource-constrained and privacy-sensitive edge computing environments. To fully realize the potential of EI services, there are several key challenges to overcome. First of all, the EI platforms should be heterogeneity-compatible. In the future, there are many dispersive EI service providers/vendors, and the common open standard should be set such that users can enjoy seamless and smooth services across heterogeneous EI platforms anywhere and anytime. Second, there are many AI programming frameworks available (e.g., Tensorflow, Torch, and

Caffe). In the future, the portability of the edge AI models trained by different programming frameworks across heterogeneously distributed edge nodes should be supported. Third, there are many programming frameworks designed specifically for edge devices (e.g., TensorFlow Lite, Caffe2, CoreML, and MXNet), however, empirical measurements [Zhang et al., 2018d] show that there is no single winner that can outperform other frameworks in all metrics. A framework that performs efficiently on more metrics can be expected in the future. Last but not least, lightweight virtualization and computing techniques such as container and function computing should be further explored to enable efficient EI service placement and migration over resource-constrained edge environments.

8.3.4   SMART SERVICE AND INCENTIVE MODELS FOR EI

The EI ecosystem will be a grand open consortium that consists of EI service providers and users, which can include but not limited to: platform providers (e.g., Amazon), AI software providers (e.g., SenseTime), edge device providers (e.g., Hikvision), network operators (e.g., AT&T), data generators (e.g., IoT and mobile device owners), and service consumers (i.e., EI users). The high-efficiency operation of EI services may require close collaboration and integration across different service providers, e.g., for implementing expanded resource sharing and smooth service handover. Thus, proper incentive mechanisms and business models are essential to stimulate effective and efficient cooperation among all members of EI ecosystem. Also, for EI service, a user can be a service consumer and meanwhile a data generator as well. In this case, a novel smart pricing scheme is needed to factorize user's service consumption and the value of its data contribution.

As a means for decentralized collaboration, blockchain with a smart contract may be integrated into EI service by running on decentralized edge servers. It is worthwhile to do research on how to smartly charge the price and properly distribute the revenue among the members in the EI ecosystem according to their proof of work. Also, designing resource-friendly lightweight blockchain consensus protocol for edge intelligence is highly desirable.

By the distributed nature of edge computing, edge devices and nodes that offer EI functionality are dispersive across diverse geo-locations and regions. Different edge devices and nodes may run different AI models and deploy different specific AI tasks. Therefore, it is important to design efficient service discovery protocols such that users can identify and locate the relevant EI service providers to fulfill their need in a timely manner. Also, to fully exploit the dispersive resource across edge nodes and devices, the partition of complex edge AI models into small subtasks and efficiently offloading these tasks among the edge nodes and devices for collaborative executions are essential.

On one hand, we need to take into account both communication and computation resource constraints on model partition and offloading. On the other hand, we should also consider the different characteristics between model training and inference tasks. For instance, model training generally involves forward and backward message passing for gradient updates, and is more delay-tolerant. How to jointly optimize both EI model training and inference task of-floading will be an important and challenging topic.

Since for many EI application scenarios (e.g., smart cities), the service environments are of high dynamics and it is hard to accurately predict future events. As a result, it would require the outstanding capability of online edge resource orchestration and provisioning to continuously accommodate massive EI tasks. Real-time joint optimization of heterogeneous computation, communication, and cache resource allocations and the high-dimensional system parameter configuration (e.g., choosing the proper model training and inference techniques) tailored to diverse task demands is critical. To tackle the algorithm design complexity, an emerging research direction is to leverage the AI techniques such as deep reinforcement learning to adapt efficient resource allocation policy in a data-driven self-learning way.

Bibliography

S. S. Abadeh, P. M. M. Esfahani, and D. Kuhn. 2015. Distributionally robust logistic regression. In *Advances in Neural Information Processing Systems*, pages 1576–1584. 90, 92, 93

M. Abadi, P. Barham, J. Chen, Z. Chen, A. Davis, J. Dean, M. Devin, S. Ghemawat, G. Irving, M. Isard, et al. 2016. Tensorflow: A system for large-scale machine learning. In *OSDI*, 16:265–283. 44, 123

R. P. Adams and D. J. MacKay. 2007. Bayesian online changepoint detection. *ArXiv:0710.3742*. 161

Z. Akhtar, Y. S. Nam, R. Govindan, S. G. Rao, J. Chen, E. Katz-Bassett, B. Ribeiro, J. Zhan, and H. Zhang. 2018. Oboe: Auto-tuning video ABR algorithms to network conditions. In *Proc. of the Conference of the ACM Special Interest Group on Data Communication, SIGCOMM*, Budapest, Hungary, August 20–25, 2018. DOI: 10.1145/3230543.3230558 159, 162, 164

L. Ambrosio and N. Gigli. 2013. A user's guide to optimal transport. In *Modelling and Optimisation of Flows on Networks*, pages 1–155, Springer. DOI: 10.1007/978-3-642-32160-3_1 88

M. M. Amiri and D. Gunduz. 2019. Machine learning at the wireless edge: Distributed stochastic gradient descent over-the-air. *ArXiv:1901.00844*. DOI: 10.1109/TSP.2020.2981904 29, 36

G. Ananthanarayanan, P. Bahl, P. Bodík, K. Chintalapudi, M. Philipose, L. Ravindranath, and S. Sinha. 2017. Real-time video analytics: The killer a page for edge computing. *computer*, 50(10):58–67. 3, 20, 169, 171

A. Antoniou, H. Edwards, and A. Storkey. 2018. How to train your MAML. *ArXiv Preprint ArXiv:1810.09502*. 46

A. Asadi, Q. Wang, and V. Mancuso. 2014. A survey on device-to-device communication in cellular networks. *IEEE Communications Surveys and Tutorials*, 16(4):1801–1819. DOI: 10.1109/comst.2014.2319555 15

J. Bergstra, O. Breuleux, F. Bastien, P. Lamblin, R. Pascanu, G. Desjardins, J. Turian, D. Warde-Farley, and Y. Bengio. 2010. Theano: A CPU and GPU math expression compiler. In *Proc. of the Python for Scientific Computing Conference (SciPy)*, vol. 4, Austin, TX. 123

J. Blanchet and K. Murthy. 2019. Quantifying distributional model risk via optimal transport. *Mathematics of Operations Research*, 44(2):565–600. DOI: 10.1287/moor.2018.0936 69

M. Blot, D. Picard, M. Cord, and N. Thome. 2016. Gossip training for deep learning. *ArXiv:1611.09726*. 29, 41, 42, 49

T. Bolukbasi, J. Wang, O. Dekel, and V. Saligrama. 2017. Adaptive neural networks for efficient inference. *ArXiv:1702.07811*. 133, 140

K. Bonawitz, V. Ivanov, B. Kreuter, A. Marcedone, H. B. McMahan, S. Patel, D. Ramage, A. Segal, and K. Seth. 2017. Practical secure aggregation for privacy-preserving machine learning. In *Proc. of the ACM SIGSAC Conference on Computer and Communications Security*, pages 1175–1191. DOI: 10.1145/3133956.3133982 32

F. Bonomi, R. Milito, J. Zhu, and S. Addepalli. 2012. Fog computing and its role in the internet of things. In *Proc. of the 1st Edition of the MCC Workshop on Mobile Cloud Computing*, pages 13–16, ACM. DOI: 10.1145/2342509.2342513 12

P. Bosilj, T. Duckett, and G. Cielniak. 2018. Analysis of morphology-based features for classification of crop and weeds in precision agriculture. *IEEE Robotics and Automation Letters*, page 1. DOI: 10.1109/lra.2018.2848305 173

L. Bottou. 2010. Large-scale machine learning with stochastic gradient descent. In *Proc. of COMPSTAT*, pages 177–186, Springer. DOI: 10.1007/978-3-7908-2604-3_16 7, 111

S. Boyd, A. Ghosh, B. Prabhakar, and D. Shah. 2006. Randomized gossip algorithms. *IEEE Transactions on Information Theory*, 52(6):2508–2530. DOI: 10.1109/tit.2006.874516 29, 41

C. Canel, T. Kim, G. Zhou, C. Li, H. Lim, D. G. Andersen, M. Kaminsky, and S. R. Dulloor. February 2018. Picking interesting frames in streaming video. *Proc. of the Conference on Systems and Machine Learning (SysML'18)*, pages 1–3, Stanford, CA. 133, 142

L. Chang, C. Yu, L. Yan, G. Chen, V. Vokkarane, Y. Ma, S. Chen, and H. Peng. 2017. A new deep learning-based food recognition system for dietary assessment on an edge computing service infrastructure. *IEEE Transactions on Services Computing*, 11(2):249–261. DOI: 10.1109/tsc.2017.2662008 174

Y. Chauvin and D. E. Rumelhart. 2013. *Backpropagation: Theory, Architectures, and Applications*, Psychology Press. DOI: 10.4324/9780203763247 7

C. Chelba and A. Acero. 2006. Adaptation of maximum entropy capitalizer: Little data can help a lot. *Computer Speech and Language*, 20(4):382–399. DOI: 10.1016/j.csl.2005.05.005 45

F. Chen, Z. Dong, Z. Li, and X. He. 2018a. Federated meta-learning for recommendation. *ArXiv Preprint ArXiv:1802.07876.* 54

Q. Chen, Z. Zheng, C. Hu, D. Wang, and F. Liu. 2019. Data-driven task allocation for multi-task transfer learning on the edge. In *IEEE 39th International Conference on Distributed Computing Systems (ICDCS).* DOI: 10.1109/icdcs.2019.00107 29, 39

T. Chen, M. Li, Y. Li, M. Lin, N. Wang, M. Wang, T. Xiao, B. Xu, C. Zhang, and Z. Zhang. 2015a. MXNet: A flexible and efficient machine learning library for heterogeneous distributed systems. *ArXiv Preprint ArXiv:1512.01274.* 123

T. Y.-H. Chen, L. Ravindranath, S. Deng, P. Bahl, and H. Balakrishnan. 2015b. Glimpse: Continuous, real-time object recognition on mobile devices. In *Proc. of the 13th ACM Conference on Embedded Networked Sensor Systems,* pages 155–168. DOI: 10.1145/2809695.2809711 133, 140

X. Chen, L. Pu, L. Gao, W. Wu, and D. Wu. 2017a. Exploiting massive D2D collaboration for energy-efficient mobile edge computing. *IEEE Wireless Communications,* 24(4):64–71. DOI: 10.1109/mwc.2017.1600321 2, 15

X. Chen, Q. Shi, L. Yang, and J. Xu. 2018b. Thriftyedge: Resource-efficient edge computing for intelligent IoT applications. *IEEE Network,* 32(1):61–65. DOI: 10.1109/mnet.2018.1700145 152

Y. Chen, Q. Feng, and W. Shi. 2018c. An industrial robot system based on edge computing: An early experience. In *USENIX Workshop on Hot Topics in Edge Computing (HotEdge).* 174

Y.-H. Chen, J. Emer, and V. Sze. 2016. Eyeriss: A spatial architecture for energy-efficient dataflow for convolutional neural networks. In *ACM SIGARCH Computer Architecture News,* 44:367–379. DOI: 10.1145/3007787.3001177 133, 134

Z. Chen, L. Jiang, W. Hu, K. Ha, B. Amos, P. Pillai, A. G. Hauptmann, and M. Satyanarayanan. 2015c. Early implementation experience with wearable cognitive assistance applications. In *Proc. of the Workshop on Wearable Systems and Applications, WearSys,* pages 33–38, Florence, Italy, May 18. DOI: 10.1145/2753509.2753517 171

Z. Chen, W. Hu, J. Wang, S. Zhao, B. Amos, G. Wu, K. Ha, K. Elgazzar, P. Pillai, R. L. Klatzky, D. P. Siewiorek, and M. Satyanarayanan. 2017b. An empirical study of latency in an emerging class of edge computing applications for wearable cognitive assistance. In *Proc. of the 2nd ACM/IEEE Symposium on Edge Computing (SEC),* pages 14:1–14:14. DOI: 10.1145/3132211.3134458 171

E. Chung, J. Fowers, K. Ovtcharov, M. Papamichael, and D. Burger. 2018. Serving DNNs in real time at datacenter scale with project brainwave. *IEEE Micro,* 38(2):8–20. DOI: 10.1109/mm.2018.022071131 135

192 BIBLIOGRAPHY

Cisco, 2016. Cisco Global Cloud Index: Forecast and Methodology, 2016–2021 White Paper. https://www.cisco.com/c/en/us/solutions/collateral/service-provider/global-cloud-index-gci/white-paper-c11-738085.html 1, 20

Cisco, 2017. Cisco Visual Networking Index: Forecast and Trends, 2017–2022. https://www.cisco.com/c/en/us/solutions/collateral/service-provider/visual-networking-index-vni/white-paper-c11-741490.html 17

R. Collobert, J. Weston, L. Bottou, M. Karlen, K. Kavukcuoglu, and P. Kuksa. 2011. Natural language processing (almost) from scratch. *Journal of Machine Learning Research*, 12(Aug):2493–2537. 6

M. Courbariaux, Y. Bengio, and J. P. David. 2014. Training deep neural networks with low precision multiplications. *ArXiv:1412.7024*. 135

M. Courbariaux, I. Hubara, D. Soudry, E. Y. Ran, and Y. Bengio. 2016. Binarized neural networks: Training deep neural networks with weights and activations constrained to +1 or -1. *ArXiv: 1602.02830*. 135

E. Cuervo, A. Balasubramanian, D.-K. Cho, A. Wolman, S. Saroiu, R. Chandra, and P. Bahl. 2010. Maui: Making smartphones last longer with code offload. In *Proc. of the 8th International Conference on Mobile Systems, Applications, and Services*, pages 49–62, ACM. DOI: 10.1145/1814433.1814441 12

J. Daily, A. Vishnu, C. Siegel, T. Warfel, and V. Amatya. 2018. GossipGraD: Scalable deep learning using gossip communication based asynchronous gradient descent. *ArXiv:1803.05880*. 29, 42, 50

L. Deng, D. Yu, et al. 2014. Deep learning: Methods and applications. *Foundations and Trends® in Signal Processing*, 7(3–4):197–387. DOI: 10.1561/2000000039 1

Y. Deng, M. M. Kamani, and M. Mahdavi. 2020. Adaptive personalized federated learning. *ArXiv Preprint ArXiv:2003.13461*. 31

A. Devarakonda, M. Naumov, and M. Garland. 2017. Adabatch: Adaptive batch sizes for training deep neural networks. *ArXiv Preprint ArXiv:1712.02029*. 117

J. Dilley, B. Maggs, J. Parikh, H. Prokop, R. Sitaraman, and B. Weihl. 2002. Globally distributed content delivery. *IEEE Internet Computing*, 6(5):50–58. DOI: 10.1109/mic.2002.1036038 11

C. T. Dinh, N. H. Tran, and T. D. Nguyen. 2020. Personalized federated learning with Moreau envelopes. *ArXiv Preprint ArXiv:2006.08848*. 31

H. T. Dinh, C. Lee, D. Niyato, and P. Wang. 2013. A survey of mobile cloud computing: Architecture, applications, and approaches. *Wireless Communications and Mobile Computing*, 13(18):1587–1611. DOI: 10.1002/wcm.1203 12

U. Drolia, K. Guo, and P. Narasimhan. 2017a. Precog: Prefetching for image recognition applications at the edge. In *Proc. of the ACM/IEEE Symposium on Edge Computing (SEC)*. 133, 141

U. Drolia, K. Guo, J. Tan, R. Gandhi, and P. Narasimhan. 2017b. Cachier: Edge-caching for recognition applications. In *Proc. of IEEE ICDCS*. DOI: 10.1109/icdcs.2017.94 133, 141, 148

M. Du, K. Wang, Y. Chen, X. Wang, and Y. Sun. 2018. Big data privacy preserving in multi-access edge computing for heterogeneous internet of things. *IEEE Communications Magazine*, 56(8):62–67. DOI: 10.1109/mcom.2018.1701148 185

Edge Computing. https://www.microsoft.com/en-us/research/project/edge-computing/ 15

R. Edmunds, N. Golmant, V. Ramasesh, P. Kuznetsov, P. Patil, and R. Puri. 2017. Transferability of adversarial attacks in model-agnostic meta-learning. *Deep Learning and Security Workshop (DLSW)*, Singapore. 54, 68

A. E. Eshratifar, M. S. Abrishami, and M. Pedram. 2018. JointDNN: An efficient training and inference engine for intelligent mobile cloud computing services. *ArXiv Preprint ArXiv:1801.08618*. DOI: 10.1109/tmc.2019.2947893 109, 110, 111, 123

T. Evgeniou and M. Pontil. 2004. Regularized multi–task learning. In *Proc. of the 10th ACM SIGKDD International Conference on Knowledge Discovery and Data Mining*, pages 109–117. DOI: 10.1145/1014052.1014067 55

A. Fallah, A. Mokhtari, and A. Ozdaglar. 2019. On the convergence theory of gradient-based model-agnostic meta-learning algorithms. *ArXiv Preprint ArXiv:1908.10400*. 30, 54

B. Fang, X. Zeng, and M. Zhang. 2018a. NestDNN: Resource-aware multi-tenant on-device deep learning for continuous mobile vision. In *Proc. of ACM Mobicom*. DOI: 10.1145/3241539.3241559 133, 144

Z. Fang, M. Luo, T. Yu, O. J. Mengshoel, M. B. Srivastava, and R. K. Gupta. 2018b. Mitigating multi-tenant interference in continuous mobile offloading. In *International Conference on Cloud Computing*, pages 20–36, Springer. DOI: 10.1007/978-3-319-94295-7_2 133, 145

C. Finn, P. Abbeel, and S. Levine. 2017. Model-agnostic meta-learning for fast adaptation of deep networks. In *Proc. of the 34th International Conference on Machine Learning*, 70:1126–1135. JMLR.org 30, 46, 53, 54

194   BIBLIOGRAPHY

C. Finn, A. Rajeswaran, S. Kakade, and S. Levine. 2019. Online meta-learning. *ArXiv Preprint ArXiv:1902.08438.* 60

J. Flinn and M. Satyanarayanan. 1999. Energy-aware adaptation for mobile applications, vol. 33, *ACM*. DOI: 10.1145/319344.319155 11

N. Fournier and A. Guillin. 2015. On the rate of convergence in Wasserstein distance of the empirical measure. *Probability Theory and Related Fields*, 162(3–4):707–738. DOI: 10.1007/s00440-014-0583-7 87

J. Fowers, M. Ghandi, S. Heil, P. Patel, A. Sapek, G. Weisz, L. Woods, S. Lanka, S. K. Reinhardt, and A. M. Caulfield. 2018. A configurable cloud-scale DNN processor for real-time AI. In *ACM/IEEE International Symposium on Computer Architecture (ISCA)*. DOI: 10.1109/isca.2018.00012 135

J. Gao, W. Fan, J. Jiang, and J. Han. 2008. Knowledge transfer via multiple model local structure mapping. In *Proc. of the 14th ACM SIGKDD International Conference on Knowledge Discovery and Data Mining*, pages 283–291. DOI: 10.1145/1401890.1401928 30, 45

R. Gao and A. J. Kleywegt. 2016. Distributionally robust stochastic optimization with Wasserstein distance. *ArXiv Preprint ArXiv:1604.02199.* 98

Gartner, 2018. 5 Trends Emerge in the Gartner Hype Cycle for Emerging Technologies, 2018. https://www.gartner.com/smarterwithgartner/5-trends-emerge-in-gartner-hype-cycle-for-emerging-technologies-2018/ 3

P. Georgiev, N. D. Lane, K. K. Rachuri, and C. Mascolo. 2016. Leo: Scheduling sensor inference algorithms across heterogeneous mobile processors and network resources. In *Proc. of the 22nd Annual International Conference on Mobile Computing and Networking*, pages 320–333, ACM. DOI: 10.1145/2973750.2973777 138

GitHub, 2013. Bayesian changepoint detection. https://github.com/hildensia/bayesian_changepoint_detection 164

GitHub, 2018. Edge Computing Landscape. https://github.com/State-of-the-Edge/landscape 175

A. Go, R. Bhayani, and L. Huang. 2009. Twitter sentiment classification using distant supervision. *CS224N Project Report*, 1(12), Stanford. 74

I. Goodfellow, J. Pouget-Abadie, M. Mirza, B. Xu, D. Warde-Farley, S. Ozair, A. Courville, and Y. Bengio. 2014a. Generative adversarial nets. In *Advances in Neural Information Processing Systems*, pages 2672–2680. 9

I. J. Goodfellow, J. Shlens, and C. Szegedy. 2014b. Explaining and harnessing adversarial examples. *ArXiv Preprint ArXiv:1412.6572*. 77

M. Gowda, A. Dhekne, S. Shen, R. R. Choudhury, L. Yang, S. Golwalkar, and A. Essanian. 2017. Bringing IoT to sports analytics. In *14th USENIX Symposium on Networked Systems Design and Implementation (NSDI)*, pages 499–513. 174

P. Goyal, P. Dollár, R. Girshick, P. Noordhuis, L. Wesolowski, A. Kyrola, A. Tulloch, Y. Jia, and K. He. 2017. Accurate, large minibatch SGD: Training ImageNet in 1 hour. *ArXiv Preprint ArXiv:1706.02677*. 111

E. Grant, C. Finn, S. Levine, T. Darrell, and T. Griffiths. 2018. Recasting gradient-based meta-learning as hierarchical Bayes. *ArXiv Preprint ArXiv:1801.08930*. 46

M. Grant, S. Boyd, and Y. Ye, 2008. CVX: Matlab software for disciplined convex programming. 91

P. M. Grulich and F. Nawab. 2018. Collaborative edge and cloud neural networks for real-time video processing. *Proc. of the VLDB Endowment*, 11(12):2046–2049. DOI: 10.14778/3229863.3236256 146

P. Guo, B. Hu, R. Li, and W. Hu. 2018. FoggyCache: Cross-device approximate computation reuse. In *Proc. of ACM Mobicom*. DOI: 10.1145/3241539.3241557 133, 141, 148, 171

A. Gupta, B. Eysenbach, C. Finn, and S. Levine. 2018. Unsupervised meta-learning for reinforcement learning. *ArXiv Preprint ArXiv:1806.04640*. 46

K. Ha, P. Pillai, G. Lewis, S. Simanta, S. Clinch, N. Davies, and M. Satyanarayanan. 2013. The impact of mobile multimedia applications on data center consolidation. In *Proc. of IEEE IC2E*. DOI: 10.1109/ic2e.2013.17 16, 17

K. Ha, Z. Chen, W. Hu, W. Richter, P. Pillai, and M. Satyanarayanan. 2014. Towards wearable cognitive assistance. In *Proc. of the 12th Annual International Conference on Mobile Systems, Applications, and Services*, pages 68–81, ACM. DOI: 10.1145/2594368.2594383 3, 12, 171

S. Han, H. Mao, and W. J. Dally. 2015a. Deep compression: Compressing deep neural networks with pruning, trained quantization and Huffman coding. *ArXiv Preprint ArXiv:1510.00149*. 133, 134, 136

S. Han, J. Pool, J. Tran, and W. Dally. 2015b. Learning both weights and connections for efficient neural network. In *Advances in Neural Information Processing Systems*, pages 1135–1143. 133, 134

A. Harlap, D. Narayanan, A. Phanishayee, V. Seshadri, G. R. Ganger, and P. B. Gibbons. February 2018. PipeDream: Pipeline parallelism for DNN training. *Proc. of the Conference on Systems and Machine Learning (SysML'18)*, pages 1–3, Stanford, CA. 29, 38, 49

K. He, X. Zhang, S. Ren, and J. Sun. 2016. Deep residual learning for image recognition. In *Computer Vision and Pattern Recognition*, pages 770–778. DOI: 10.1109/cvpr.2016.90 6, 8

Y. He, G. J. Mendis, and J. Wei. 2017. Real-time detection of false data injection attacks in smart grid: A deep learning-based intelligent mechanism. *IEEE Transactions on Smart Grid*, 8(5):2505–2516. DOI: 10.1109/tsg.2017.2703842 174

Y. He, J. Lin, Z. Liu, H. Wang, L.-J. Li, and S. Han. 2018. AMC: Automl for model compression and acceleration on mobile devices. In *European Conference on Computer Vision*, pages 815–832, Springer. DOI: 10.1007/978-3-030-01234-2_48 180

B. Heintz, A. Chandra, and R. K. Sitaraman. 2015. Optimizing grouped aggregation in geo-distributed streaming analytics. In *Proc. of ACM HPDC*. DOI: 10.1145/2749246.2749276 1

G. Hinton, O. Vinyals, and J. Dean. 2015. Distilling the knowledge in a neural network. *ArXiv Preprint ArXiv:1503.02531*. 45

S. Hochreiter and J. Schmidhuber. 1997. Long short-term memory. *Neural Computation*, 9(8):1735–1780. DOI: 10.1162/neco.1997.9.8.1735 9

M. S. Hossain, G. Muhammad, and S. U. Amin. 2018. Improving consumer satisfaction in smart cities using edge computing and caching: A case study of date fruits classification. *Future Generation Computer Systems*, 88:333–341. DOI: 10.1016/j.future.2018.05.050 174

A. G. Howard, M. Zhu, B. Chen, D. Kalenichenko, W. Wang, T. Weyand, M. Andreetto, and H. Adam. 2017. MobileNets: Efficient convolutional neural networks for mobile vision applications. *ArXiv Preprint ArXiv:1704.04861*. 6, 136

K. Hsieh, A. Harlap, N. Vijaykumar, D. Konomis, G. R. Ganger, P. B. Gibbons, and O. Mutlu. 2017. Gaia: Geo-distributed machine learning approaching {LAN} speeds. In *14th {USENIX} Symposium on Networked Systems Design and Implementation (NSDI)*, pages 629–647. 29, 33, 48

Y. C. Hu, M. Patel, D. Sabella, N. Sprecher, and V. Young. 2015. Mobile edge computing—a key technology towards 5G. *ETSI White Paper*, 11(11):1–16. 15

C. Huang, W. Bao, D. Wang, and F. Liu. 2019. Dynamic adaptive DNN surgery for inference acceleration on the edge. In *Proc. of IEEE INFOCOM*. DOI: 10.1109/infocom.2019.8737614 133, 137

J. Huang, V. Rathod, C. Sun, M. Zhu, A. Korattikara, A. Fathi, I. Fischer, Z. Wojna, Y. Song, S. Guadarrama, and K. Murphy. July 2017a. Speed/accuracy trade-offs for modern convolutional object detectors. In *The IEEE Conference on Computer Vision and Pattern Recognition (CVPR)*. DOI: 10.1109/cvpr.2017.351 182

R. H. Huang, B. J. Chang, Y. L. Tsai, and Y. H. Liang. 2017b. Mobile edge computing-based vehicular cloud of cooperative adaptive driving for platooning autonomous self driving. In *IEEE 7th International Symposium on Cloud and Service Computing (SC2)*. DOI: 10.1109/sc2.2017.13 174

Y. Huang, L. Chu, Z. Zhou, L. Wang, J. Liu, J. Pei, and Y. Zhang. 2020. Personalized federated learning: An attentive collaboration approach. *ArXiv Preprint ArXiv:2007.03797*. 31

C.-C. Hung, G. Ananthanarayanan, P. Bodik, L. Golubchik, M. Yu, P. Bahl, and M. Philipose. 2018. VideoEdge: Processing camera streams using hierarchical clusters. In *IEEE/ACM Symposium on Edge Computing (SEC)*, pages 115–131. DOI: 10.1109/sec.2018.00016 133, 145, 147, 169

I. IHS Markit, 2018. Top Video Surveillance Trends for 2018. https://ihsmarkit.com/Info/1217/top-video-surveillance-trends-2018.html 169

J. Dean and U. Hölzle. 2019. Build and train machine learning models on our new Google cloud TPUs. https://www.blog.google/products/google-cloud/google-cloud-offer-tpus-machine-learning/ 44

S. Jain, J. Jiang, Y. Shu, G. Ananthanarayanan, and J. Gonzalez. 2018. Rexcam: Resource-efficient, cross-camera video analytics at enterprise scale. *ArXiv:1811.01268*. 133, 142

V. Jalaparti, I. Bliznets, S. Kandula, B. Lucier, and I. Menache. 2016. Dynamic pricing and traffic engineering for timely inter-datacenter transfers. In *Proc. of ACM SIGCOMM*. DOI: 10.1145/2934872.2934893 16

E. Jeong, S. Oh, H. Kim, J. Park, M. Bennis, and S.-L. Kim. 2018a. Communication-efficient on-device machine learning: Federated distillation and augmentation under non-IID private data. *ArXiv Preprint ArXiv:1811.11479*. 30, 45

H.-J. Jeong, H.-J. Lee, C. H. Shin, and S.-M. Moon. 2018b. IONN: Incremental offloading of neural network computations from mobile devices to edge servers. In *Proc. of the ACM Symposium on Cloud Computing*, pages 401–411. DOI: 10.1145/3267809.3267828 133, 138

A. H. Jiang, D. L.-K. Wong, C. Canel, L. Tang, I. Misra, M. Kaminsky, M. A. Kozuch, P. Pillai, D. G. Andersen, and G. R. Ganger. 2018a. Mainstream: Dynamic stem-sharing for multi-tenant video processing. In *Proc. of USENIX ATC*. 133, 144, 169

J. Jiang, G. Ananthanarayanan, P. Bodik, S. Sen, and I. Stoica. 2018b. Chameleon: Scalable adaptation of video analytics. In *Proc. of ACM SIGCOMM*. DOI: 10.1145/3230543.3230574 133, 143, 145, 146, 147, 169

C. Jie, X. Lanyu, R. Abdallah, and S. Weisong. 2017a. An OS for internet-of-everything: Early experience from a smart home prototype. *Zte Communications*, (4):12–22. 172

C. Jie, L. Xu, R. Abdallah, and W. Shi. 2017b. Edgeos_h: A home operating system for internet of everything. In *Proc. of IEEE ICDCS*. DOI: 10.1109/icdcs.2017.325 3, 173

P. H. Jin, Q. Yuan, F. Iandola, and K. Keutzer. 2016. How to scale distributed deep learning? *ArXiv Preprint ArXiv:1611.04581.* 29, 41, 42, 50

N. P. Jouppi, A. Borchers, R. Boyle, P. L. Cantin, and B. Nan. June 26, 2017. In-datacenter performance analysis of a tensor processing unit. *Proc. of the 44th Annual International Symposium on Computer Architecture*, pages 1–12, ACM, Toronto, Canada. DOI: 10.1145/3140659.3080246 135

D. Kang, J. Emmons, F. Abuzaid, P. Bailis, and M. Zaharia. 2017a. NoScope: Optimizing neural network queries over video at scale. *Proc. of the VLDB Endowment*, 10(11):1586–1597. DOI: 10.14778/3137628.3137664 133, 141, 148

Y. Kang, J. Hauswald, C. Gao, A. Rovinski, T. Mudge, J. Mars, and L. Tang. 2017b. Neurosurgeon: Collaborative intelligence between the cloud and mobile edge. *ACM SIGARCH Computer Architecture News*, 45(1):615–629. DOI: 10.1145/3093337.3037698 133, 136, 147

L. V. Kantorovich. 2006. On the translocation of masses. *Journal of Mathematical Sciences*, 133(4):1381–1382. DOI: 10.1007/s10958-006-0049-2 88

A. Karbalayghareh, X. Qian, and E. R. Dougherty. 2018. Optimal Bayesian transfer learning. *IEEE Transactions on Signal Processing*, 66(14):3724–3739. DOI: 10.1109/tsp.2018.2839583 83

H. Kim, J. Park, M. Bennis, and S.-L. Kim. 2018. On-device federated learning via blockchain and its latency analysis. *ArXiv:1808.03949.* 29, 32, 48

A. J. Kleywegt, A. Shapiro, and T. Homem-de Mello. 2002. The sample average approximation method for stochastic discrete optimization. *SIAM Journal on Optimization*, 12(2):479–502. DOI: 10.1137/s1052623499363220 92, 94

J. H. Ko, T. Na, M. F. Amir, and S. Mukhopadhyay. 2018. Edge-host partitioning of deep neural networks with feature space encoding for resource-constrained internet-of-things platforms. *ArXiv:1802.03835.* DOI: 10.1109/avss.2018.8639121 133, 137

J. Konečný, H. B. McMahan, F. X. Yu, P. Richtárik, A. T. Suresh, and D. Bacon. 2016. Federated learning: Strategies for improving communication efficiency. *ArXiv Preprint ArXiv:1610.05492.* 29, 32

A. Krizhevsky, G. Hinton, et al. 2009. Learning multiple layers of features from tiny images. *Online Technical Report*, University of Toronto, pages 1–60. https://www.cs.toronto.edu/~kriz/learning-features-2009-TR.pdf 122, 163

A. Krizhevsky, I. Sutskever, and G. E. Hinton. 2012. ImageNet classification with deep convolutional neural networks. In *Proc. of NIPS*. DOI: 10.1145/3065386 6, 8, 122, 152, 162

J. F. Kurose and K. W. Ross. 2009. *Computer Networking: A Top-Down Approach*, vol. 3, Addison Wesley Boston. 16

H. A. Lagar-Cavilla, N. Tolia, E. De Lara, M. Satyanarayanan, and D. O'Hallaron. 2007. Interactive resource-intensive applications made easy. In *ACM/IFIP/USENIX International Conference on Distributed Systems Platforms and Open Distributed Processing*, pages 143–163, Springer. DOI: 10.1007/978-3-540-76778-7_8 12

A. Lalitha, S. Shekhar, T. Javidi, and F. Koushanfar. 2018. Fully decentralized federated learning. In *3rd Workshop on Bayesian Deep Learning (NeurIPS)*, pages 1–6, Montreal, Canada. 29, 32

N. Lane, S. Bhattacharya, P. Georgiev, C. Forlivesi, L. Jiao, L. Qendro, and F. Kawsar. 2016. DeepX: A software accelerator for low-power deep learning inference on mobile devices. In *International Conference on Information Processing in Sensor Networks*, page 23. DOI: 10.1109/ipsn.2016.7460664 133, 138, 147

Y. LeCun, L. Bottou, Y. Bengio, P. Haffner, et al. 1998. Gradient-based learning applied to document recognition. *Proc. of the IEEE*, 86(11):2278–2324. DOI: 10.1109/5.726791 74, 93, 107, 122

Y. Lecun, Y. Bengio, and G. Hinton. 2015. Deep learning. *Nature*, 521(7553):436. DOI: 10.1038/nature14539 1

S. Leroux, S. Bohez, E. De Coninck, T. Verbelen, B. Vankeirsbilck, P. Simoens, and B. Dhoedt. 2017. The cascading neural network: Building the internet of smart things. *Knowledge and Information Systems*, 52(3):791–814. DOI: 10.1007/s10115-017-1029-1 133, 140

D. Li, Z. Zhang, W. Liao, and Z. Xu. 2018a. KLRA: A kernel level resource auditing tool for IoT operating system security. In *IEEE/ACM Symposium on Edge Computing (SEC)*, pages 427–432. DOI: 10.1109/sec.2018.00058 185

E. Li, Z. Zhou, and X. Chen. 2018b. Edge intelligence: On-demand deep learning model co-inference with device-edge synergy. In *Proc. of the Workshop on Mobile Edge Communications (MECOMM)*, pages 31–36. DOI: 10.1145/3229556.3229562 3, 133, 139, 147

H. Li, C. Hu, J. Jiang, Z. Wang, Y. Wen, and W. Zhu. 2018c. JALAD: Joint accuracy- and latency-aware deep structure decoupling for edge-cloud execution. In *IEEE 24th International Conference on Parallel and Distributed Systems (ICPADS)*, pages 671–678. DOI: 10.1109/padsw.2018.8645013 111, 123, 133, 137, 148

L. Li, K. Ota, and M. Dong. 2018d. Deep learning for smart industry: Efficient manufacture inspection system with fog computing. *IEEE Transactions on Industrial Informatics*. DOI: 10.1109/tii.2018.2842821 3, 133, 140, 147, 172

X. Li and J. Bilmes. 2007. A Bayesian divergence prior for classifier adaptation. In *Artificial Intelligence and Statistics*, pages 275–282. 45

Y. Li, J. Park, M. Alian, Y. Yuan, Z. Qu, P. Pan, R. Wang, A. Schwing, H. Esmaeilzadeh, and N. S. Kim. 2018e. A network-centric hardware/algorithm co-design to accelerate distributed training of deep neural networks. In *51st Annual IEEE/ACM International Symposium on Microarchitecture (MICRO)*, pages 175–188. DOI: 10.1109/micro.2018.00023 37, 49

Z. Li, F. Zhou, F. Chen, and H. Li. 2017. Meta-SGD: Learning to learn quickly for few-shot learning. *ArXiv Preprint ArXiv:1707.09835*. 46

S. Lin, G. Yang, and J. Zhang, 2020. A collaborative learning framework via federated meta-learning. to be presented in *40th IEEE International Conference on Distributed Computing Systems (ICDCS)*, 2020. 30, 54

Y. Lin, S. Han, H. Mao, Y. Wang, and W. J. Dally. 2017. Deep gradient compression: Reducing the communication bandwidth for distributed training. *ArXiv Preprint ArXiv:1712.01887*. 29, 36, 48

Q. Liu, X. Liao, and L. Carin. 2008. Semi-supervised multitask learning. In *Advances in Neural Information Processing Systems*, pages 937–944. DOI: 10.1109/tpami.2008.296 96, 103

S. Liu, Y. Lin, Z. Zhou, K. Nan, H. Liu, and J. Du. 2018. On-demand deep model compression for mobile devices: A usage-driven model selection framework. In *Proc. of the 16th Annual International Conference on Mobile Systems, Applications, and Services*, pages 389–400, ACM. DOI: 10.1145/3210240.3210337 133, 136, 147

W. Liu, D. Anguelov, D. Erhan, C. Szegedy, S. Reed, C.-Y. Fu, and A. C. Berg. 2016. SSD: Single shot multibox detector. In *European Conference on Computer Vision*, pages 21–37, Springer. DOI: 10.1007/978-3-319-46448-0_2 6

C. Lo, Y.-Y. Su, C.-Y. Lee, and S.-C. Chang. 2017. A dynamic deep neural network design for efficient workload allocation in edge computing. In *Computer Design (ICCD), IEEE International Conference on*, pages 273–280. DOI: 10.1109/iccd.2017.49 133, 140

A. R. Lobo, D. K. Nadgir, and S. T. S. Kuncolienkar, January 16, 2014. Intelligent edge caching. US Patent App. 13/548,584. 18

R. Lomathe. 2012. Edge computing. *Pacific Northwest National Laboratory (PNLL)*. 13, 14

A. Luckow, M. Cook, N. Ashcraft, E. Weill, E. Djerekarov, and B. Vorster. 2017. Deep learning in the automotive industry: Applications and tools. In *IEEE International Conference on Big Data*. DOI: 10.1109/bigdata.2016.7841045 172

D. Mahajan, J. Park, E. Amaro, H. Sharma, A. Yazdanbakhsh, J. K. Kim, and H. Esmaeilzadeh. 2016. Tabla: A unified template-based framework for accelerating statistical machine learning. In *High Performance Computer Architecture (HPCA), IEEE International Symposium on*, pages 14–26. DOI: 10.1109/hpca.2016.7446050 43

B. K. Mallick and S. G. Walker. 1997. Combining information from several experiments with nonparametric priors. *Biometrika*, 84(3):697–706. DOI: 10.1093/biomet/84.3.697 95

H. Mao, S. Yao, T. Tang, B. Li, J. Yao, and Y. Wang. 2018a. Towards real-time object detection on embedded systems. *IEEE Transactions on Emerging Topics in Computing*, 6(3):417–431. DOI: 10.1109/tetc.2016.2593643 6

J. Mao, X. Chen, K. W. Nixon, C. Krieger, and Y. Chen. 2017a. MoDNN: Local distributed mobile computing system for deep neural network. In *Design, Automation and Test in Europe Conference and Exhibition (DATE)*, pages 1396–1401, IEEE. DOI: 10.23919/date.2017.7927211 133, 138

J. Mao, Z. Yang, W. Wen, C. Wu, L. Song, K. W. Nixon, X. Chen, H. Li, and Y. Chen. 2017b. MeDNN: A distributed mobile system with enhanced partition and deployment for large-scale DNNs. In *Proc. of the 36th International Conference on Computer-Aided Design*, pages 751–756, IEEE Press. DOI: 10.1109/iccad.2017.8203852 133, 138

Y. Mao, C. You, J. Zhang, K. Huang, and K. B. Letaief. 2017c. A survey on mobile edge computing: The communication perspective. *IEEE Communications Surveys and Tutorials*, 19(4):2322–2358. DOI: 10.1109/comst.2017.2745201 3, 15

Y. Mao, S. Yi, Q. Li, J. Feng, F. Xu, and S. Zhong. 2018b. A privacy-preserving deep learning approach for face recognition with edge computing. In *Proc. USENIX Workshop Hot Topics Edge Comput. (HotEdge)*, pages 1–6. 29, 37

A. Mathur, N. D. Lane, S. Bhattacharya, A. Boran, C. Forlivesi, and F. Kawsar. 2017. DeepEye: Resource efficient local execution of multiple deep vision models using wearable commodity hardware. In *Proc. of ACM Mobisys*. DOI: 10.1145/3081333.3081359 133, 145

V. Mathur and K. Chahal. 2018. Hydra: A peer to peer distributed training and data collection framework. *ArXiv Preprint ArXiv:1811.09878.* 109

H. B. McMahan, E. Moore, D. Ramage, S. Hampson, et al. 2016. Communication-efficient learning of deep networks from decentralized data. *ArXiv Preprint ArXiv:1602.05629.* 28, 29, 30, 32, 48, 55, 74

P. Micikevicius, S. Narang, J. Alben, G. Diamos, E. Elsen, D. Garcia, B. Ginsburg, M. Houston, O. Kuchaiev, and G. Venkatesh. 2018. Mixed precision training. *ArXiv: 1710.03740.* 135

Microsoft, 2016. Democratizing AI. https://news.microsoft.com/features/democratizing-ai/ 20

M. Mirza and S. Osindero. 2014. Conditional generative adversarial nets. *ArXiv Preprint ArXiv:1411.1784.* 93, 107

N. Mishra, M. Rohaninejad, X. Chen, and P. Abbeel. 2017. A simple neural attentive meta-learner. *ArXiv Preprint ArXiv:1707.03141.* 30, 46

D. Miyashita, E. H. Lee, and B. Murmann. 2016. Convolutional neural networks using logarithmic data representation. *ArXiv Preprint ArXiv:1603.01025.* 135

V. Mnih, K. Kavukcuoglu, D. Silver, A. A. Rusu, J. Veness, M. G. Bellemare, A. Graves, M. Riedmiller, A. K. Fidjeland, G. Ostrovski, et al. 2015. Human-level control through deep reinforcement learning. *Nature*, 518(7540):529. DOI: 10.1038/nature14236 10

A. G. Mohapatra, S. K. Lenka, and B. Keswani. 2018. Neural network and fuzzy logic based smart DSS model for irrigation notification and control in precision agriculture. *Proc. of the National Academy of Sciences India*, (5):1–10. DOI: 10.1007/s40010-017-0401-6 173

S. Mukhopadhyay and A. E. Gelfand. 1997. Dirichlet process mixed generalized linear models. *Journal of the American Statistical Association*, 92(438):633–639. DOI: 10.1080/01621459.1997.10474014 95

V. Mulhollon. 2004. Wondershaper. http://manpages.ubuntu.com/manpages/trusty/man8/ wondershaper.8.html 162

P. Müller, F. Quintana, and G. Rosner. 2004. A method for combining inference across related nonparametric Bayesian models. *Journal of the Royal Statistical Society: Series B (Statistical Methodology)*, 66(3):735–749. DOI: 10.1111/j.1467-9868.2004.05564.x 95

T. Munkhdalai, X. Yuan, S. Mehri, and A. Trischler. 2017. Rapid adaptation with conditionally shifted neurons. *ArXiv Preprint ArXiv:1712.09926.* 46

M. Murshed, C. Murphy, D. Hou, N. Khan, G. Ananthanarayanan, and F. Hussain. 2019. Machine learning at the network edge: A survey. *ArXiv Preprint ArXiv:1908.00080.* 111

A. Nagabandi, I. Clavera, S. Liu, R. S. Fearing, P. Abbeel, S. Levine, and C. Finn. 2018. Learning to adapt in dynamic, real-world environments through meta-reinforcement learning. *ArXiv Preprint ArXiv:1803.11347.* 46

D. Narayanan, K. Santhanam, A. Phanishayee, and M. Zaharia. 2018. Accelerating deep learning workloads through efficient multi-model execution. In *NIPS Workshop on Systems for Machine Learning*. 133, 144

A. Nichol, J. Achiam, and J. Schulman. 2018. On first-order meta-learning algorithms. *ArXiv Preprint ArXiv:1803.02999*. 46

T. Nishio and R. Yonetani. 2018. Client selection for federated learning with heterogeneous resources in mobile edge. *ArXiv:1804.08333*. DOI: 10.1109/icc.2019.8761315 29, 35

Y. H. Oh, Q. Quan, D. Kim, S. Kim, J. Heo, S. Jung, J. Jang, and J. W. Lee. 2018. A portable, automatic data quantizer for deep neural networks. In *Proc. of ACM PACT*. 133, 134

S. Omidshafiei, J. Pazis, C. Amato, J. P. How, and J. Vian. 2017. Deep decentralized multi-task multi-agent reinforcement learning under partial observability. In *Proc. of the 34th International Conference on Machine Learning*, 70:2681–2690. JMLR.org 30, 45

S. A. Osia, A. S. Shamsabadi, A. Taheri, K. Katevas, S. Sajadmanesh, H. R. Rabiee, N. D. Lane, and H. Haddadi. 2017. A hybrid deep learning architecture for privacy-preserving mobile analytics. *ArXiv:1703.02952*. DOI: 10.1109/jiot.2020.2967734 29, 37, 41

E. Park, D. Kim, S. Kim, Y.-D. Kim, G. Kim, S. Yoon, and S. Yoo. 2015. Big/little deep neural network for ultra low power inference. In *Proc. of the 10th International Conference on Hardware/Software Codesign and System Synthesis*, pages 124–132, IEEE Press. DOI: 10.1109/codesisss.2015.7331375 133, 143

J. Park, H. Sharma, D. Mahajan, J. K. Kim, P. Olds, and H. Esmaeilzadeh. 2017. Scale-out acceleration for machine learning. In *Proc. of the 50th Annual IEEE/ACM International Symposium on Microarchitecture*, pages 367–381. DOI: 10.1145/3123939.3123979 43

A. Paszke, S. Gross, S. Chintala, G. Chanan, E. Yang, Z. DeVito, Z. Lin, A. Desmaison, L. Antiga, and A. Lerer. 2017. Automatic differentiation in pytorch. In *Proc. 31st Conference on Neural Information Processing Systems (NIPS)*, pages 1–4, Long Beach, CA. 123

J. Pennington, R. Socher, and C. Manning. 2014. Glove: Global vectors for word representation. In *Proc. of the Conference on Empirical Methods in Natural Language Processing (EMNLP)*, pages 1532–1543. DOI: 10.3115/v1/d14-1162 74

C. Perera, C. H. L. Member, S. Jayawardena, and M. Chen. 2015. Context-aware computing in the internet of things: A survey on internet of things from industrial market perspective. *ArXiv:1502.00164*. 18

Q. Pu, G. Ananthanarayanan, P. Bodik, S. Kandula, A. Akella, P. Bahl, and I. Stoica. 2015. Low latency geo-distributed data analytics. In *Proc. of ACM SIGCOMM*. DOI: 10.1145/2829988.2787505 1

J. Qiu, S. Song, W. Yu, H. Yang, W. Jie, Y. Song, K. Guo, B. Li, E. Zhou, and J. Yu. 2016. Going deeper with embedded FPGA platform for convolutional neural network. In *ACM/SIGDA International Symposium on Field-programmable Gate Arrays (FPGA)*. DOI: 10.1145/2847263.2847265 135

Z. Quan, Q. Zhang, W. Shi, and Z. Hong. 2018. Firework: Data processing and sharing for hybrid cloud-edge analytics. *IEEE Transactions on Parallel and Distributed Systems (TPDS)*, PP(99):1. DOI: 10.1109/tpds.2018.2812177 169

Q. Quanto. 2016. Datasets Over Algorithms. https://www.kdnuggets.com/2016/05/datasets-over-algorithms.html/ 20

K. Rakelly, A. Zhou, D. Quillen, C. Finn, and S. Levine. 2019. Efficient off-policy meta-reinforcement learning via probabilistic context variables. *ArXiv Preprint ArXiv:1903.08254.* 46

X. Ran, H. Chen, X. Zhu, Z. Liu, and J. Chen. 2018. DeepDecision: A mobile deep learning framework for edge video analytics. In *Proc. of IEEE INFOCOM*. DOI: 10.1109/infocom.2018.8485905 133, 146, 148

M. Rapp, R. Khalili, and J. Henkel. 2020. Distributed learning on heterogeneous resource-constrained devices. *ArXiv Preprint ArXiv:2006.05403.* 32

S. Ravi and H. Larochelle. 2017. Optimization as a model for few-shot learning. In *International Conference on Learning Representations.* 46, 53

B. Reagen, P. Whatmough, R. Adolf, S. Rama, H. Lee, S. K. Lee, J. M. Hernández-Lobato, G.-Y. Wei, and D. Brooks. 2016. Minerva: Enabling low-power, highly-accurate deep neural network accelerators. In *ACM/IEEE 43rd Annual International Symposium on Computer Architecture (ISCA)*, pages 267–278. DOI: 10.1109/isca.2016.32 133, 136, 147

J. Redmon, S. Divvala, R. Girshick, and A. Farhadi. 2016. You only look once: Unified, real-time object detection. In *Proc. of the IEEE conference on computer vision and pattern recognition*, pages 779–788. DOI: 10.1109/cvpr.2016.91 6

J. Ren, G. Yu, and G. Ding. 2019. Accelerating DNN training in wireless federated edge learning system. *ArXiv Preprint ArXiv:1905.09712.* 109, 110

S. Ruder. 2017. An overview of multi-task learning in deep neural networks. *ArXiv Preprint ArXiv:1706.05098.* 55

D. E. Rumelhart, G. E. Hinton, and R. J. Williams. 1986. Learning representations by back-propagating errors. *Nature*, 323(6088):533. DOI: 10.1038/323533a0 7

I. Sa, Z. Ge, F. Dayoub, B. Upcroft, T. Perez, and C. McCool. 2016. DeepFruits: A fruit detection system using deep neural networks. *Sensors*, 16(8):1222. DOI: 10.3390/s16081222 173

A. K. Sahu, T. Li, M. Sanjabi, M. Zaheer, A. Talwalkar, and V. Smith. 2018. On the convergence of federated optimization in heterogeneous networks. *ArXiv Preprint ArXiv:1812.06127*. 55, 74

M. Satyanarayanan, P. Bahl, R. Cáceres, and N. Davies. 2009. The case for VM-based cloudlets in mobile computing. *IEEE Pervasive Computing*, 8(4):14–23. DOI: 10.1109/mprv.2009.82 12, 20

M. Satyanarayanan et al. 2001. Pervasive computing: Vision and challenges. *IEEE Personal Communications*, 8(4):10–17. DOI: 10.1109/98.943998 11

J. Schmidhuber. 1987. Evolutionary principles in self-referential learning. Ph.D. thesis, Technische Universität München. 53

S. Seneviratne, Y. Hu, T. Nguyen, G. Lan, S. Khalifa, K. Thilakarathna, M. Hassan, and A. Seneviratne. 2017. A survey of wearable devices and challenges. *IEEE Communications Surveys and Tutorials*, 19(4):1. DOI: 10.1109/comst.2017.2731979 17

S. Shafieezadeh-Abadeh, D. Kuhn, and P. M. Esfahani. 2017. Regularization via mass transportation. *ArXiv Preprint ArXiv:1710.10016*. 97, 98

H. Shao, H. Jiang, H. Zhao, and F. Wang. 2016. An enhancement deep feature fusion method for rotating machinery fault diagnosis. *Knowledge-Based Systems*, 119:200–220. DOI: 10.1016/j.knosys.2016.12.012 172

A. Shapiro, D. Dentcheva, and A. Ruszczyński. 2009. Lectures on stochastic programming: Modeling and theory. *SIAM*. DOI: 10.1137/1.9780898718751 68

R. Sharma, S. Biookaghazadeh, B. Li, and M. Zhao. 2018. Are existing knowledge transfer techniques effective for deep learning with edge devices? In *IEEE International Conference on Edge Computing (EDGE)*, pages 42–49. DOI: 10.1109/edge.2018.00013 39

W. Shi, J. Cao, Q. Zhang, Y. Li, and L. Xu. 2016. Edge computing: Vision and challenges. *IEEE Internet of Things Journal*, 3(5):637–646. DOI: 10.1109/jiot.2016.2579198 2, 3, 14, 15, 18

R. Shokri and V. Shmatikov. 2015. Privacy-preserving deep learning. In *Proc. of the 22nd ACM SIGSAC Conference on Computer and Communications Security*, pages 1310–1321, ACM. DOI: 10.1145/2810103.2813687 29, 30, 48

G. Shu, W. Liu, X. Zheng, and J. Li. 2018. If-CNN: Image-aware inference framework for CNN with the collaboration of mobile devices and cloud. *IEEE Access*, 6:621–633. DOI: 10.1109/access.2018.2880196 133, 143

K. Simonyan and A. Zisserman. 2014. Very deep convolutional networks for large-scale image recognition. *ArXiv:1409.1556*. 6, 8

A. Sinha, H. Namkoong, and J. Duchi. 2017. Certifying some distributional robustness with principled adversarial training. *ArXiv Preprint ArXiv:1710.10571*. 69, 70, 72

V. Smith, C.-K. Chiang, M. Sanjabi, and A. S. Talwalkar. 2017. Federated multi-task learning. In *Advances in Neural Information Processing Systems*, pages 4424–4434. 53, 55

J. Snell, K. Swersky, and R. Zemel. 2017. Prototypical networks for few-shot learning. In *Advances in Neural Information Processing Systems*, pages 4077–4087. 30, 46

S. Srinivas and F. Fleuret. 2018. Knowledge transfer with Jacobian matching. *ArXiv Preprint ArXiv:1803.00443*. 46

D. Stamoulis, T.-W. R. Chin, A. K. Prakash, H. Fang, S. Sajja, M. Bognar, and D. Marculescu. 2018. Designing adaptive neural networks for energy-constrained image classification. In *Proc. of ACM ICCAD*. DOI: 10.1145/3240765.3240796 133, 144

S. U. Stich, J.-B. Cordonnier, and M. Jaggi. 2018. Sparsified SGD with memory. In *Advances in Neural Information Processing Systems*, pages 4452–4463. 29, 36

I. Stoica, D. Song, R. A. Popa, D. Patterson, M. W. Mahoney, R. Katz, A. D. Joseph, M. Jordan, J. M. Hellerstein, J. E. Gonzalez, et al. 2017. A Berkeley view of systems challenges for AI. *ArXiv:1712.05855*. 21

D. Svozil, V. Kvasnicka, and J. Pospichal. 1997. Introduction to multi-layer feed-forward neural networks. *Chemometrics and Intelligent Laboratory Systems*, 39(1):43–62. DOI: 10.1016/s0169-7439(97)00061-0 4

V. Sze, Y.-H. Chen, T.-J. Yang, and J. S. Emer. 2017. Efficient processing of deep neural networks: A tutorial and survey. *Proc. of the IEEE*, 105(12):2295–2329. DOI: 10.1109/jproc.2017.2761740 42, 146

C. Szegedy, W. Liu, Y. Jia, P. Sermanet, S. Reed, D. Anguelov, D. Erhan, V. Vanhoucke, and A. Rabinovich. 2015. Going deeper with convolutions. In *Proc. of the IEEE Conference on Computer Vision and Pattern Recognition*, pages 1–9. DOI: 10.1109/cvpr.2015.7298594 8

R. Tandon, Q. Lei, A. G. Dimakis, and N. Karampatziakis. 2017. Gradient coding: Avoiding stragglers in distributed learning. In *International Conference on Machine Learning*, pages 3368–3376. 183

H. Tang, S. Gan, C. Zhang, T. Zhang, and J. Liu. 2018. Communication compression for decentralized training. In *Advances in Neural Information Processing Systems*, pages 7663–7673. 29, 36

Z. Tao and Q. Li. 2018. ESGD: Communication efficient distributed deep learning on the edge. In *USENIX Workshop on Hot Topics in Edge Computing (HotEdge 18)*, Boston, MA. 29, 36, 49

B. Taylor, V. S. Marco, W. Wolff, Y. Elkhatib, and Z. Wang. 2018. Adaptive deep learning model selection on embedded systems. In *Proc. of ACM LCTES*. DOI: 10.1145/3211332.3211336 133, 143, 148

S. Teerapittayanon, B. McDanel, and H. Kung. 2016. BranchyNet: Fast inference via early exiting from deep neural networks. In *Pattern Recognition (ICPR), 23rd International Conference on*, pages 2464–2469, IEEE. DOI: 10.1109/icpr.2016.7900006 133, 139, 154, 162

S. Teerapittayanon, B. McDanel, and H. Kung. 2017. Distributed deep neural networks over the cloud, the edge and end devices. In *Distributed Computing Systems (ICDCS), IEEE 37th International Conference on*, pages 328–339. DOI: 10.1109/icdcs.2017.226 133, 139, 148

Y. W. Teh. 2010. Dirichlet process. *Encyclopedia of Machine Learning*, pages 280–287. DOI: 10.1007/978-1-4899-7687-1_219 95

S. Thrun and L. Pratt. 2012. *Learning to Learn*, Springer Science and Business Media. DOI: 10.1007/978-1-4615-5529-2_1 46

S. Tokui, K. Oono, S. Hido, and J. Clayton. 2015. Chainer: A next-generation open source framework for deep learning. In *Proc. of Workshop on Machine Learning Systems (LearningSys) in the 29th Annual Conference on Neural Information Processing Systems (NIPS)*, 5:1–6. 123, 162

N. Tolia, D. G. Andersen, and M. Satyanarayanan. 2006. Quantifying interactive user experience on thin clients. *Computer*, (3):46–52. DOI: 10.1109/mc.2006.101 12

D. Tsipras, S. Santurkar, L. Engstrom, A. Turner, and A. Madry. 2018. Robustness may be at odds with accuracy. *ArXiv Preprint ArXiv:1805.12152*. 68

J. van der Hooft, S. Petrangeli, T. Wauters, R. Huysegems, P. R. Alface, T. Bostoen, and F. De Turck. 2016. HTTP/2-based adaptive streaming of HEVC video over 4G/LTE networks. *IEEE Communications Letters*, 20(11):2177–2180. DOI: 10.1109/lcomm.2016.2601087 162, 165

D. Vasisht, Z. Kapetanovic, J. Won, X. Jin, R. Chandra, S. N. Sinha, A. Kapoor, M. Sudarshan, and S. Stratman. 2017. Farmbeats: An IoT platform for data-driven agriculture. In *14th*

USENIX Symposium on Networked Systems Design and Implementation (NSDI), pages 515–529. 173

S. Venkataramani, A. Ranjan, S. Banerjee, D. Das, S. Avancha, A. Jagannathan, A. Durg, D. Nagaraj, B. Kaul, P. Dubey, et al. 2017. ScaleDeep: A scalable compute architecture for learning and evaluating deep networks. In *ACM SIGARCH Computer Architecture News*, 45:13–26. 42

S. Venugopal, M. Gazzetti, Y. Gkoufas, and K. Katrinis. 2018. Shadow puppets: Cloud-level accurate {AI} inference at the speed and economy of edge. In *{USENIX} Workshop on Hot Topics in Edge Computing (HotEdge 18)*. 133, 141

C. Villani. 2008. *Optimal Transport: Old and New*, vol. 338, Springer Science and Business Media. 86

O. Vinyals, C. Blundell, T. Lillicrap, D. Wierstra, et al. 2016. Matching networks for one shot learning. In *Advances in Neural Information Processing Systems*, pages 3630–3638. 46

R. Viswanathan, G. Ananthanarayanan, and A. Akella. 2016. Clarinet: Wan-aware optimization for analytics queries. In *Proc. of USENIX OSDI*. 17

R. Volpi, H. Namkoong, O. Sener, J. C. Duchi, V. Murino, and S. Savarese. 2018. Generalizing to unseen domains via adversarial data augmentation. In *Advances in Neural Information Processing Systems*, pages 5334–5344. 70

A. Vulimiri, C. Curino, B. Godfrey, J. Padhye, and G. Varghese. 2015. Global analytics in the face of bandwidth and regulatory constraints. In *Proc. of USENIX NSDI*. 17

J. Wang, Z. Feng, Z. Chen, S. George, M. Bala, P. Pillai, S.-W. Yang, and M. Satyanarayanan. 2018a. Bandwidth-efficient live video analytics for drones via edge computing. In *IEEE/ACM Symposium on Edge Computing (SEC)*, pages 159–173. DOI: 10.1109/sec.2018.00019 133, 141

J. Wang, J. Zhang, W. Bao, X. Zhu, B. Cao, and P. S. Yu. 2018b. Not just privacy: Improving performance of private deep learning in mobile cloud. In *Proc. of the 24th ACM SIGKDD International Conference on Knowledge Discovery and Data Mining*, pages 2407–2416. DOI: 10.1145/3219819.3220106 29, 37, 41, 49

Q. Wang, Y. Li, and P. Li. 2016. Liquid state machine based pattern recognition on FPGA with firing-activity dependent power gating and approximate computing. In *Circuits and Systems (ISCAS), IEEE International Symposium on*, pages 361–364. DOI: 10.1109/iscas.2016.7527245 42, 43

Q. Wang, Y. Li, B. Shao, S. Dey, and P. Li. 2017a. Energy efficient parallel neuromorphic architectures with approximate arithmetic on FPGA. *Neurocomputing*, 221:146–158. DOI: 10.1016/j.neucom.2016.09.071 43

S. Wang, X. Zhang, Y. Zhang, L. Wang, J. Yang, and W. Wang. 2017b. A survey on mobile edge networks: Convergence of computing, caching and communications. *IEEE Access*, 5:6757–6779. DOI: 10.1109/access.2017.2685434 15

S. Wang, T. Tuor, T. Salonidis, K. K. Leung, C. Makaya, T. He, and K. Chan. 2019a. Adaptive federated learning in resource constrained edge computing systems. *IEEE Journal on Selected Areas in Communications*, 37(6):1205–1221. DOI: 10.1109/jsac.2019.2904348 29, 33, 55, 63

X. Wang, M. Chen, T. Taleb, A. Ksentini, and V. Leung. 2014. Cache in the air: Exploiting content caching and delivery techniques for 5G systems. *IEEE Communications Magazine*, 52(2):131–139. DOI: 10.1109/mcom.2014.6736753 18

X. Wang, Y. Han, C. Wang, Q. Zhao, X. Chen, and M. Chen. 2019b. In-edge AI: Intelligentizing mobile edge computing, caching and communication by federated learning. *IEEE Network*. DOI: 10.1109/mnet.2019.1800286 3

P. J. Werbos. 1990. Backpropagation through time: What it does and how to do it. *Proc. of the IEEE*, 78(10):1550–1560. DOI: 10.1109/5.58337 9

C. Wu, D. Brooks, K. Chen, D. Chen, S. Choudhury, M. Dukhan, K. Hazelwood, E. Isaac, Y. Jia, B. Jia, T. Leyvand, H. Lu, Y. Lu, L. Qiao, B. Reagen, J. Spisak, F. Sun, A. Tulloch, P. Vajda, X. Wang, Y. Wang, B. Wasti, Y. Wu, R. Xian, S. Yoo, and P. Zhang. February 2019. Machine learning at Facebook: Understanding inference at the edge. In *IEEE International Symposium on High Performance Computer Architecture (HPCA)*, pages 331–344. DOI: 10.1109/hpca.2019.00048 132

C. J. Wu et al. 1983. On the convergence properties of the em algorithm. *The Annals of Statistics*, 11(1):95–103. DOI: 10.1214/aos/1176346060 102

Z. Xu, L. Chao, and X. Peng. 2018. T-rest: An open-enabled architectural style for the internet of things. *IEEE Internet-of-Things Journal*. DOI: 10.1109/jiot.2018.2875912 172

Z. Xu, Z. Yang, J. Xiong, J. Yang, and X. Chen. 2019. Elfish: Resource-aware federated learning on heterogeneous edge devices. *ArXiv Preprint ArXiv:1912.01684*. 32

Z.-W. Xu. 2014. Cloud-sea computing systems: Towards thousand-fold improvement in performance per watt for the coming zettabyte era. *Journal of Computer Science and Technology*, 29(2):177–181. DOI: 10.1007/s11390-014-1420-2 13

T.-J. Yang, Y.-H. Chen, and V. Sze. 2017. Designing energy-efficient convolutional neural networks using energy-aware pruning. In *IEEE Conference on Computer Vision and Pattern Recognition (CVPR)*. DOI: 10.1109/cvpr.2017.643 133, 134

S. Yi, Z. Hao, Z. Qin, and Q. Li. 2015. Fog computing: Platform and applications. In *3rd IEEE Workshop on Hot Topics in Web Systems and Technologies (HotWeb)*, pages 73–78. DOI: 10.1109/hotweb.2015.22 16

S. Yi, Z. Hao, Q. Zhang, Z. Quan, and Q. Li. 2017. Lavea: Latency-aware video analytics on edge computing platform. In *IEEE International Conference on Distributed Computing Systems (ICDCS)*. DOI: 10.1109/icdcs.2017.182 169

C. Yin, J. Tang, Z. Xu, and Y. Wang. 2018. Adversarial meta-learning. *ArXiv Preprint ArXiv:1806.03316*. 54, 68

J. Yosinski, J. Clune, Y. Bengio, and H. Lipson. 2014. How transferable are features in deep neural networks? In *Advances in Neural Information Processing Systems*, pages 3320–3328. 39

Y. You, I. Gitman, and B. Ginsburg. 2017. Large batch training of convolutional networks. *ArXiv Preprint ArXiv:1708.03888*. 111

L. Zeng, E. Li, Z. Zhou, and X. Chen. 2019. Boomerang: On-demand cooperative deep neural network inference for edge intelligence on industrial internet of things. *IEEE Network*. DOI: 10.1109/mnet.001.1800506 133

C. Zhang, Q. Cao, H. Jiang, W. Zhang, J. Li, and J. Yao. 2018a. FFS-VA: A fast filtering system for large-scale video analytics. In *Proc. of ACM ICPP*. DOI: 10.1145/3225058.3225103 133, 142, 148

H. Zhang, G. Ananthanarayanan, P. Bodik, M. Philipose, P. Bahl, and M. J. Freedman. 2017a. Live video analytics at scale with approximation and delay-tolerance. In *Proc. of USENIX NSDI*. 146

Q. Zhang, Y. Wang, X. Zhang, L. Liu, X. Wu, W. Shi, and H. Zhong. 2018b. OpenVDAP: An open vehicular data analytics platform for cavs. In *Distributed Computing Systems (ICDCS), IEEE 38th International Conference on*. DOI: 10.1109/icdcs.2018.00131 138, 174

Q. Zhang, Q. Zhang, W. Shi, and H. Zhong. 2018c. Distributed collaborative execution on the edges and its application to amber alerts. *IEEE Internet of Things Journal*. DOI: 10.1109/jiot.2018.2845898 169, 171

X. Zhang, Y. Wang, L. Chao, C. Li, L. Wu, X. Peng, and Z. Xu. 2017b. Iehouse: A non-intrusive household appliance state recognition system. In *IEEE Smart-World/SCALCOM/UIC/ATC/CBDCom/IOP/SCI*. DOI: 10.1109/uic-atc.2017.8397510 173

X. Zhang, Y. Wang, and W. Shi. 2018d. PCAMP: Performance comparison of machine learning packages on the edges. In *{USENIX} Workshop on Hot Topics in Edge Computing (HotEdge 18)*. 186

Z. Zhang, Y. Chen, and J. Zhang. 2020a. Distributionally robust learning based on Dirichlet process prior in edge networks. *Journal of Communications and Information Networks*, 5(1):26–39. 81

Z. Zhang, Y. Chen, and J. Zhang. 2020b. Distributionally robust edge learning with Dirichlet process prior. to be presented in *40th IEEE International Conference on Distributed Computing Systems (ICDCS)*. 81

Z. Zhang, S. Lin, M. Dedeoglu, K. Ding, and J. Zhang. 2020c. Data-driven distributionally robust optimization for edge intelligence. In *39th IEEE Conference on Computer Communications, INFOCOM*, pages 2619–2628, Toronto, ON, Canada, July 6–9. DOI: 10.1109/infocom41043.2020.9155440 81

J. Zhao, R. Mortier, J. Crowcroft, and L. Wang. 2018a. Privacy-preserving machine learning based data analytics on edge devices. DOI: 10.1145/3278721.3278778 32, 48

Z. Zhao, K. M. Barijough, and A. Gerstlauer. 2018b. DeepThings: Distributed adaptive deep learning inference on resource-constrained IoT edge clusters. *IEEE Transactions on Computer-Aided Design of Integrated Circuits and Systems*, 37(11):2348–2359. DOI: 10.1109/tcad.2018.2858384 133, 138

Z. Zhou, S. Yang, L. J. Pu, and S. Yu. 2020. CEFL: Online admission control, data scheduling and accuracy tuning for cost-efficient federated learning across edge nodes. *IEEE Internet of Things Journal*. DOI: 10.1109/jiot.2020.2984332 27

G. Zhu, Y. Wang, and K. Huang. 2018. Low-latency broadband analog aggregation for federated edge learning. *ArXiv:1812.11494*. 185

B. Zoph and Q. V. Le. 2016. Neural architecture search with reinforcement learning. *ArXiv:1611.01578*. 180

Authors' Biographies

SEN LIN

Sen Lin received his B.Eng. degree in Electrical Engineering from Zhejiang University, Hangzhou, China, in 2013, and his M.S. degree in Telecommunications from The Hong Kong University of Science and Technology, Hong Kong, in 2014. Currently, he is pursuing a Ph.D. degree at the School of Electrical, Computer, and Energy Engineering at Arizona State University, Tempe, AZ, USA. His current research interests include statistical machine learning, reinforcement learning, and edge computing.

ZHI ZHOU

Zhi Zhou received B.S., M.E., and Ph.D. degrees in 2012, 2014, and 2017, respectively, all from the School of Computer Science and Technology at Huazhong University of Science and Technology (HUST), Wuhan, China. He is currently a research fellow in the School of Data and Computer Science at Sun Yat-sen University, Guangzhou, China. In 2016, he was a Visiting Scholar at University of Göttingen. He was nominated for the 2019 CCF Outstanding Doctoral Dissertation Award, the sole recipient of the 2018 ACM Wuhan & Hubei Computer Society Doctoral Dissertation Award, and a recipient of the Best Paper Award of IEEE UIC 2018. His research interests include edge computing, cloud computing, and distributed systems.

ZHAOFENG ZHANG

Zhaofeng Zhang received his B.Eng. degree in Electrical Engineering from Huazhong University of Science and Technology, Wuhan, China, in 2015, and his M.S. degree in Electrical Engineering from Arizona State University, Tempe, AZ, USA, in 2017. Currently, he is pursuing a Ph.D. degree at the School of Electrical, Computer, and Energy Engineering in Arizona State University, Tempe, AZ, USA. His current research interests include edge computing, statistical machine learning, and optimization.

XU CHEN

Xu Chen received a Ph.D. degree in Information Engineering from the Chinese University of Hong Kong, in 2012. He is a Full Professor with Sun Yat-sen University, Guangzhou, China, and the Vice Director of the National and Local Joint Engineering Laboratory of Digital Home Interactive Applications. He was a Post-Doctoral Research Associate with Arizona State University, Tempe, USA, from 2012–2014, and a Humboldt Scholar Fellow with the Institute of Computer Science at the University of Göttingen, Germany, from 2014–2016. He was a recipient of the Prestigious Humboldt Research Fellowship awarded by the Alexander von Humboldt Foundation of Germany, the 2014 Hong Kong Young Scientist Runner-Up Award, the 2017 IEEE Communication Society Asia–Pacific Outstanding Young Researcher Award, the 2017 IEEE ComSoc Young Professional Best Paper Award, the Honorable Mention Award at the 2010 IEEE international conference on Intelligence and Security Informatics, the Best Paper Runner-Up Award at the 2014 IEEE International Conference on Computer Communications (INFOCOM), and the Best Paper Award at the 2017 IEEE International Conference on Communications. He is currently an Area Editor at the *IEEE Open Journal of the Communications Society*, an Associate Editor of the *IEEE Transactions Wireless Communications*, *IEEE Internet of Things Journal*, and *IEEE Journal on Selected Areas in Communications (JSAC)* Series on Network Softwarization and Enablers.

JUNSHAN ZHANG

Junshan Zhang received his Ph.D. degree from the School of ECE at Purdue University, in 2000. He joined the School of ECEE at Arizona State University in August 2000 and has been Fulton Chair Professor there since 2015. His research interests fall in the general field of information networks and data science, including communication networks, edge computing and machine learning for IoT, mobile social networks, and smart grid. His current research focuses on fundamental problems in information networks and data science, including edge computing and machine learning in IoT and 5G, IoT data privacy/security, information theory, stochastic modeling, and control for smart grid.

Prof. Zhang is a Fellow of the IEEE and a recipient of the ONR Young Investigator Award in 2005 and the NSF CAREER award in 2003. He received the IEEE Wireless Communication Technical Committee Recognition Award in 2016. His papers have won a few awards, including the Best Student Paper Award at WiOPT 2018, the Kenneth C. Sevcik Outstanding Student Paper Award at ACM SIGMETRICS/IFIP Performance 2016, the Best Paper Runner-up Award at IEEE INFOCOM 2009 and IEEE INFOCOM 2014, and the Best Paper Award at IEEE ICC 2008 and ICC 2017. Building on his research findings, he co-founded Smartiply Inc, a Fog Computing startup company delivering boosted network connectivity and embedded artificial intelligence.

Prof. Zhang was TPC co-chair for a number of major conferences in communication networks, including IEEE INFOCOM 2012 and ACM MOBIHOC 2015. He was the general chair for ACM/IEEE SEC 2017, WiOPT 2016, and IEEE Communication Theory Workshop 2007. He was a Distinguished Lecturer of the IEEE Communications Society. He is currently serving as Editor-in-chief for *IEEE Transactions on Wireless Communications* and a senior editor for *IEEE/ACM Transactions on Networking*.

Printed in the United States
by Baker & Taylor Publisher Services